T0269243

# Advances in Intelligent Systems and Computing

Volume 423

**Series editor**

Janusz Kacprzyk, Polish Academy of Sciences, Warsaw, Poland
e-mail: kacprzyk@ibspan.waw.pl

## About this Series

The series "Advances in Intelligent Systems and Computing" contains publications on theory, applications, and design methods of Intelligent Systems and Intelligent Computing. Virtually all disciplines such as engineering, natural sciences, computer and information science, ICT, economics, business, e-commerce, environment, healthcare, life science are covered. The list of topics spans all the areas of modern intelligent systems and computing.

The publications within "Advances in Intelligent Systems and Computing" are primarily textbooks and proceedings of important conferences, symposia and congresses. They cover significant recent developments in the field, both of a foundational and applicable character. An important characteristic feature of the series is the short publication time and world-wide distribution. This permits a rapid and broad dissemination of research results.

## Advisory Board

Chairman

Nikhil R. Pal, Indian Statistical Institute, Kolkata, India
e-mail: nikhil@isical.ac.in

Members

Rafael Bello, Universidad Central "Marta Abreu" de Las Villas, Santa Clara, Cuba
e-mail: rbellop@uclv.edu.cu

Emilio S. Corchado, University of Salamanca, Salamanca, Spain
e-mail: escorchado@usal.es

Hani Hagras, University of Essex, Colchester, UK
e-mail: hani@essex.ac.uk

László T. Kóczy, Széchenyi István University, Győr, Hungary
e-mail: koczy@sze.hu

Vladik Kreinovich, University of Texas at El Paso, El Paso, USA
e-mail: vladik@utep.edu

Chin-Teng Lin, National Chiao Tung University, Hsinchu, Taiwan
e-mail: ctlin@mail.nctu.edu.tw

Jie Lu, University of Technology, Sydney, Australia
e-mail: Jie.Lu@uts.edu.au

Patricia Melin, Tijuana Institute of Technology, Tijuana, Mexico
e-mail: epmelin@hafsamx.org

Nadia Nedjah, State University of Rio de Janeiro, Rio de Janeiro, Brazil
e-mail: nadia@eng.uerj.br

Ngoc Thanh Nguyen, Wroclaw University of Technology, Wroclaw, Poland
e-mail: Ngoc-Thanh.Nguyen@pwr.edu.pl

Jun Wang, The Chinese University of Hong Kong, Shatin, Hong Kong
e-mail: jwang@mae.cuhk.edu.hk

More information about this series at http://www.springer.com/series/11156

Vítězslav Stýskala · Dmitrii Kolosov
Václav Snášel · Taalaybek Karakeyev
Ajith Abraham
Editors

# Intelligent Systems for Computer Modelling

Proceedings of the 1st European-Middle Asian Conference on Computer Modelling 2015, EMACOM 2015

 Springer

*Editors*
Vítězslav Stýskala
Faculty of Electrical Engineering
  and Computer Science
VŠB—Technical University of Ostrava
Ostrava
Czech Republic

Dmitrii Kolosov
Department of Electrical Engineering
VŠB—Technical University of Ostrava
Ostrava
Czech Republic

Václav Snášel
Faculty of Electrical Engineering
  and Computer Science
VŠB—Technical University of Ostrava
Ostrava
Czech Republic

Taalaybek Karakeyev
Kyrgyz National University named after
  Jusup Balasagyn
Bishkek
Kyrgyzstan

Ajith Abraham
Scientific Network for Innovation
  and Research Excellence
Machine Intelligence Research Labs
Auburn
USA

ISSN 2194-5357          ISSN 2194-5365  (electronic)
Advances in Intelligent Systems and Computing
ISBN 978-3-319-27642-7     ISBN 978-3-319-27644-1  (eBook)
DOI 10.1007/978-3-319-27644-1

Library of Congress Control Number: 2015958334

This Springer imprint is published by SpringerNature
The registered company is Springer International Publishing AG Switzerland

# Preface

This volume of Advances in Intelligent Systems and Computing contains papers presented at the 1st European-Middle Asian Conference on Computer Modelling—EMACOM 2015. This international conference was conceived as a brand new scientific and social event of mutual collaboration between the VŠB—Technical University of Ostrava (Ostrava, Czech Republic) and the Kyrgyz National University named after Jusup Balasagyn (Bishkek, Kyrgyz Republic).

Considering the vivid heritage of the cooperation between the Czech and Kyrgyz Republics, which dates back to the "Interhelpo" cooperative relations from the early 20s of the twentieth century, this international conference successfully aimed at creating a new modern platform for mutual meetings and exchange of latest research ideas between leading European and Middle-Asian scientists and professionals.

EMACOM 2015 was widely supported by the officials of both countries. It was organized under the patronage of the president (Hejtman) of the Moravian-Silesian Region of the Czech Republic, Mr. Miroslav Novak, the Plenipotentiary Envoy of the Government of Kyrgyz Republic to the Issyk-Kyl Region, Mr. Emilbek Kaptagayev, and the Ministry of Education and Science of the Kyrgyz Republic.

The scientific aim of EMACOM 2015 was to present the latest development in the field of computer-aided modelling as an essential aspect of research and development of innovative systems and their applications. The conference showed that together with simulations, various modelling techniques, enabled and encouraged by the rapid development of high-performance computing platforms, are crucial for cost-efficient design, verification, and prototyping of solutions in many diverse industrial fields spanning the whole range from manufacturing, mining, machinery, and automotive industries to infrastructure planning and development, economics, energy, and modern agriculture and food industry.

EMACOM 2015 was one of the largest international scientific events held in the Kyrgyz Republic in 2015. It was hosted by a modern tourist resort, Raduga, located on shores of the beautiful lake Issyk-Kul—a true natural diamond of the Middle-Asian region. The EMACOM team expresses deep gratitude to the members of the International Program Committee, local officials and the management of the resort, the publisher of this volume, and, of course, all the authors, for their impressive efforts and contribution to the success of this conference.

August 2015                                                                 Vítězslav Stýskala
                                                                                 Dmitrii Kolosov
                                                                                 Václav Snášel
                                                                          Taalaybek Karakeyev
                                                                                 Ajith Abraham

# Organization

## Honorary Chairs

Ermek Z. Usekeyev, Kyrgyz National University Named after Jusup Balasagyn, Kyrgyzstan
Ivo Vondrak, VŠB—Technical University of Ostrava, Czech Republic

## Conference Chairs

Václav Snášel, VŠB—Technical University of Ostrava, Czech Republic
Taalaybek Karakeyev, Kyrgyz National University Named after Jusup Balasagyn, Kyrgyzstan

## Program Committee Chairs

Ajith Abraham, Machine Intelligence Research Labs, USA
Dmitry V. Kolosov, VŠB—Technical University of Ostrava, Czech Republic

## Proceedings Chair

Pavel Kromer, VŠB—Technical University of Ostrava, Czech Republic

## Conference Organizers

Vítězslav Stýskala, VŠB—Technical University of Ostrava, Czech Republic
Jana Nowakova, VŠB—Technical University of Ostrava, Czech Republic

Jan Platos, VŠB—Technical University of Ostrava, Czech Republic
Pavel Kromer, VŠB—Technical University of Ostrava, Czech Republic

## International Program Committee

Ajith Abraham, MIR Labs, USA
Gulziynat Nasyrbekovna Arkabayeva, Kyrgyz National University Named after
Jusup Balasagyn, Kyrgyzstan
Murat Satarkulovich Asanov, Kyrgyz National University Named after Jusup
Balasagyn, Kyrgyzstan
Batyigul J. Baiachorova, Kyrgyz National University Named after Jusup
Balasagyn, Kyrgyzstan
Bolot I. Baybosunov, I. Arabaev Kyrgyz State University, Kyrgyzstan
Anton Beláň, Slovak University of Technology in Bratislava, Slovakia
Gulmira Beyshekeyeva, Kyrgyz National University Named after Jusup Balasagyn,
Kyrgyzstan
Zhanybek Abdraevich Bokoev, Kyrgyz National University Named after Jusup
Balasagyn, Kyrgyzstan
Damir Sarymzakovich Bolotbayev, Kyrgyz National University Named after Jusup
Balasagyn, Kyrgyzstan
Jiri Bouchala, VŠB—Technical University of Ostrava, Czech Republic
Zhumgal Tukenovna Bugubayeva, Kyrgyz National University Named after Jusup
Balasagyn, Kyrgyzstan
Bayyshich Checheybayev, Kyrgyz National University Named after Jusup
Balasagyn, Kyrgyzstan
Kalybek Choroyev, Kyrgyz National University Named after Jusup Balasagyn,
Kyrgyzstan
Milan Dado, University of Žilina, Slovakia
Stephen Dodds, University of East London—Docklands, United Kingdom
Jiri Dvorsky, VŠB—Technical University of Ostrava, Czech Republic
Burul Almazbekovna Dzhunushalieva, Kyrgyz National University Named after
Jusup Balasagyn, Kyrgyzstan
Viktor Anatolevich Finochenko, Rostov State Transport University, Russia
Bronislav Firago, Belarusian National Technical University, Belaruss
Tarke Gaber, VŠB—Technical University of Ostrava, Czech Republic
Jiří Hammerbauer, University of West Bohemia, Czech Republic
Dusan Husek, Institute of Computer Science, Academy of Sciences of the Czech
Republic
Taalaybek Murzabekovich Imanaliev, American University of Central Asia,
Kyrgyzstan
Konrad Jackowski, Wroclaw University of Technology, Poland
František Janeček, Slovak University of Technology in Bratislava, Slovakia

Ivan Zelinka, VŠB—Technical University of Ostrava, Czech Republic
Yuri I. Zharkov, Rostov State Transport University, Russia

## Sponsoring Institutions

VŠB—Technical University of Ostrava, Czech Republic

# Contents

# Finite Element Modelling of T-Plate for Treatment of Distal Radius

K. Frydrýšek, G. Theisz, L. Bialy, L. Pliska and L. Pleva

**Abstract** Distal radius fractures are the most frequent type of injury in the upper limbs. This paper analyses a new locking compression plate for osteosynthesis. The plate, made of Ti6Al4V or stainless steel, is used for the internal fixation of distal radius fractures by open reposition. The bone (with a fracture of 23–C1 AO classification), together with the plate and angularly stable screws, was exposed to axial tension/compression and bending quasi-dynamic overloading. The material properties of bone can be described with sufficient accuracy for individual bone parts using a homogeneous isotropic material model. In order to provide an adequate description of reality, the bone model was divided into two types of osseous tissues, i.e. cortical (compact—substantia compacta) and spongy (cancellous—substantia spongiosa) tissues. Other bone models based on the theory of elastic (Winkler's) foundations were also used. Variant calculations (i.e. numerical simulations) were carried out for fused bone (i.e. successful treatment) and non-fused bone. The safety factor was evaluated with regard to the minimum yield strength of the material. The situation of maximum loading on non-fused bone is an extreme state in which the fracture fails to

K. Frydrýšek (✉) · G. Theisz
Faculty of Mechanical Engineering, VŠB–Technical University of Ostrava,
17. Listopadu 15/2172, 708 33 Ostrava, Czech Republic
e-mail: karel.frydrysek@vsb.cz

G. Theisz
e-mail: gunther.theisz@vsb.cz

L. Bialy · L. Pliska · L. Pleva
Trauma Centre, Ostrava University Hospital, 17. Listopadu 1790,
708 52 Ostrava, Czech Republic
e-mail: lubor.bialy@fno.cz

L. Pliska
e-mail: ludek.pliska@fno.cz

L. Pleva
e-mail: leopold.pleva@fno.cz

L. Bialy · L. Pliska · L. Pleva
Faculty of Medicine, University of Ostrava, Dvořákova 7,
701 03 Ostrava, Czech Republic

© Springer International Publishing Switzerland 2016
V. Styskala et al. (eds.), *Intelligent Systems for Computer Modelling*,
Advances in Intelligent Systems and Computing 423,
DOI 10.1007/978-3-319-27644-1_1

1

heal and the patient places excessive stress on the limb. From this biomechanical perspective, the analysed new locking compression plate with its angularly stable screws can be considered safe for the treatment of distal radius fractures.

**Keywords** Locking compression plate · Distal radius · Fractures · Strength analyses · FEM · Elastic foundation · Biomechanics · Bone modelling · Traumatology · Orthopaedics

# 1 Introduction

Various injuries, including distal radius fractures, have been a feature of human life since prehistoric times. Distal radius fractures rank among the most frequent types of fracture, and are often caused by falling from a considerable height and landing on the extended upper limb. This type of fracture is most common in adult patients of advanced age, in women suffering from osteoporosis, or in younger patients involved in sports activities. Distal radius fractures are a serious problem both in terms of their frequency and in terms of their possible consequences for the future functioning of the wrist. It is therefore important not to underestimate the treatment of these fractures, to identify more serious cases and to treat them properly.

The osteosynthesis of complicated radius fractures can be performed by plates, see Fig. 1, external fixators, see Fig. 2, nails and wires. However, this paper is focused on plates.

The development of high-quality angularly stable plates has enabled the growing number of unstable comminuted fractures to be treated by osteosynthesis using a locking compression plate (LCP). At the Trauma Centre of Ostrava University Hospital, 80 % of osteosynthesis cases are treated using LCPs. The basic types of plate for the distal radius are medial, dorsal and radial. Medial plates are used in around 95 % of cases. The clear preference for medial plates is due to the lower number of pre-operative and post-operative complications associated with this type. A further advantage is the lower risk of lesions in the tendon structure.

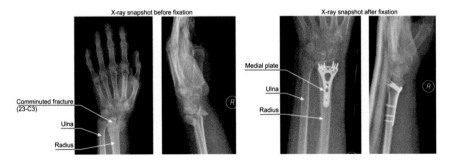

**Fig. 1** X-ray snapshots before and after fixation with plate

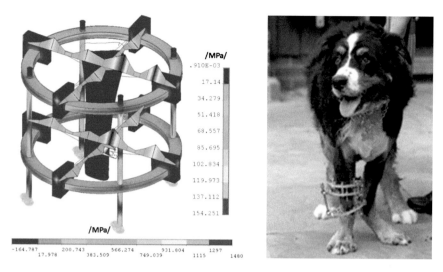

**Fig. 2** External ring fixation (an example of different possible treatment—finite element method modelling and application)

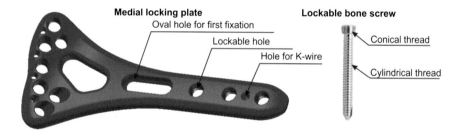

**Fig. 3** Medial plate and lockable bone screw, producer medin, a.s., Czech Republic

This article reports on a strength and deformation analysis of a new medial plate for the internal fixation of the distal radius; see Fig. 3. To fix the plate to the bone, lockable bone screws with conical and cylindrical threads are used. Simplified strength analyses of fixators are given in the literature, e.g. [1–3].

## 2 Finite Element Analysis

### 2.1 Model

The basis for the creation of an accurate mathematical model of bone is the use of X-radiation. CT (computer tomography) and MRI (magnetic resonance imaging) are able to generate series of images showing sections of a particular area of the body. For purposes of illustration, Fig. 4 shows 2 sections of the upper limb.

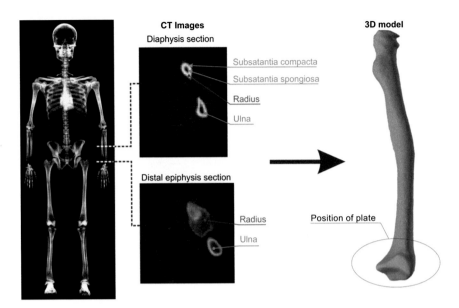

**Fig. 4** CT images of upper limb

Primarily in the area of the diaphysis there is a visible difference between compact and spongy bone tissue. The compact tissue is lighter in colour as it is denser and thus absorbs more X-radiation. The difference is less clearly visible in the area of the joints, as the long bones are generally not hollow in the area of the epiphysis, unlike in the diaphysis. To create high-quality models of bone tissue it is necessary to perform a process of selection in order to separate out the soft tissues. The complexity and difficulty of the selection depends on the quality of the CT images. In the case of poor-quality CT images, the noise must be filtered out in order to enable the separation of the soft tissues from the bone tissue. The model of the radius and the position of the plate is shown in Fig. 4. The CT images was acquired from the male patient 63 years old and processed (i.e. created CAD and FEM models via Mimics software and Ansys Workbench software).

In addition to successful treatment, a simple intra-articular fracture of the metaphysis and joint surface was also simulated, defined according to the AO classification as 23–C1. It is very difficult to accurately model a fracture in its full complexity, and currently very few data on fracture geometry are available. The fracture was therefore modelled as a 3D planar discontinuity in the bone geometry; see Fig. 4.

## 2.2 Material

Human bone tissue is a non-homogeneous and anisotropic material. It is very difficult to obtain suitable samples and determine its material properties, as these

**Table 1** Material properties of bone

|  | Elasticity modulus E /MPa/ | Poisson number μ /1/ |
| --- | --- | --- |
| Cortical tissue | 15,000 | 0.3 |
| Spongiosal tissue | 1500 | 0.3 |
| Stainless steel AISI 316L | 210,000 | 0.3 |
| Ti6Al4V | 106,000 | 0.342 |

properties depend on many factors. The dependency of mechanical properties on the direction of external loading is high. A detailed description can be given by stochastic mechanics. For this reason, practical ways of suitably modelling these complex material properties are being sought. Models of materials can then be used, depending on their quality, only up to a certain level of similarity to reality. A comparison between orthotropic, homogeneous and non-homogeneous isotropic material is given e.g. in [4] and other information about theory and experiments are written in [5]. Because this is a strength analysis of a plate and not of bone, a sufficient level of similarity to reality appears to be represented by the use of a non-homogeneous isotropic material model. This is described by the tensile elasticity modulus E and the Poisson number μ for the cortical and spongiosal tissue; see Table 1 and Fig. 5. The Poisson umber is same for all areas: $\mu = 0.3$.

The plate is produced from AISI 316L steel and Ti6Al4V titanium alloy. These materials are widely used in surgical practice. The analysis uses a homogeneous isotropic linearly flexible material model. The mechanical properties of the materials used to make the plates and lockable screws were determined by tensile testing; see Table 1.

## 2.3 Boundary Conditions

Two states of loading were considered; the implant system is subjected to tensile loading and compression loading. Tensile and compression force acts on the axis of the radius. The wrist bones are substituted by the boundary condition of an elastic

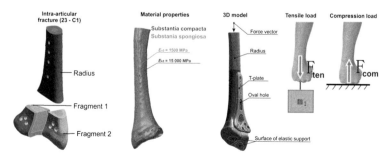

**Fig. 5** Intra-articular fracture, material properties model of radius with plate and load

foundation; see Fig. 5. The size of the surface to which this boundary condition is applied corresponds in the distal radius epiphysis to the contact surfaces with the wrist bones.

The loading tensile force $F_{ten}$ was calculated for the case of lifting a weight $m_1$; see Fig. 5 and Eq. 1. The weight was set as 20 kg. According to the literature [6], the radius transfers 80 % of loading (the ulna 20 %), so the loading coefficient $f_1 = 0.8$ was applied in the calculation of tensile force. The dynamic coefficient $f_2$ was also applied. The value of the dynamic coefficient was set at 1.5. The value of gravitational acceleration is $g = 9.807$ m s$^{-2}$.

$$F_{ten} = m_1 \times g \times f_1 \times f_2 = 20 \times 9.807 \times 0.8 \times 1.5 = 235.4\,N \qquad (1)$$

The loading compression force $F_{com}$ was calculated for the case of a person standing up after sitting on a chair, i.e. resting their body on a hand; see Fig. 5 and Eq. 2. If a person weighing 100 kg raises their body from a chair, then each upper limb bears $m_2 = 50$ kg. In this case the calculation of force involved the same loading coefficient $f_1$ and the same dynamic coefficient $f_2$ as in the previous case.

$$F_{com} = m_2 \times g \times f_1 \times f_2 = 50 \times 9.807 \times 0.8 \times 1.5 = 588.4\,N \qquad (2)$$

## 2.4 Results

Figure 6 (Example of stress analysis, Finite Element Method) shows the distribution of reduced stress $\sigma_{red}$ according to von Mises theory for a plate consisting of AISI 316L material. This calculation is performed for non-fused bone (i.e. unsuccessful treatment).

Variant calculations were carried out for fused bone and non-fused bone and for Ti6Al4V or stainless steel materials. With regard to the minimum yield strength of the material, the safety factor $k_{\sigma y}/1/$ is calculated using Eq. 3, where $\sigma_{redMAX}$/MPa/ is the maximum reduced stress according to von Mises theory and $\sigma_y$/MPa/ is the minimum yield strength of the given material. The situation of maximum loading on non-fused bone is an extreme state in which the fracture fails to heal and the patient places excessive stress on the limb.

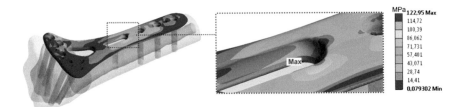

**Fig. 6** Example of stress analysis (equivalent von Mises stress /MPa/, axial tension, non-fused bone, AISI 316L material, Ansys Workbench software)

**Table 2** Results of strength analyses and safety factor for axial tension ($F_{ten}$ = 235.4 N, finite element method, Ansys Workbench software)

| | | Material | |
|---|---|---|---|
| | | Titanium (Ti6Al4V) | Stainless steel (AISI 316L) |
| Minimum material yield strength $\sigma_{redMAX}$/MPa/ | | 919 | 892 |
| Non-fused bone (unsuccessful treatment) | Maximum calculated stress $\sigma_{redMAX}$/MPa/ | 111.98 | 122.95 |
| | Safety factor $k_{\sigma y}$/1/ | **8.2** | **7.2** |
| Fused bone (successful treatment) | Maximum calculated stress $\sigma_{redMAX}$/MPa/ | 20.5 | 25.24 |
| | Safety factor $k_{\sigma y}$/1/ | **44.8** | **14.3** |

**Table 3** Results of strength analyses and safety factor for axial compression ($F_{com}$ = 588.4 N, finite element method, Ansys Workbench software)

| | | Material | |
|---|---|---|---|
| | | Titanium (Ti6Al4V) | Stainless steel (AISI 316L) |
| Minimum material yield strength $\sigma_{redMAX}$/MPa/ | | 919 | 892 |
| Non-fused bone (unsuccessful treatment) | Maximum calculated stress $\sigma_{redMAX}$/MPa/ | 51.13 | 61.62 |
| | Safety factor $k_{\sigma y}$/1/ | **17.9** | **14.5** |
| Fused bone (successful treatment) | Maximum calculated stress $\sigma_{redMAX}$/MPa/ | 51.54 | 62.53 |
| | Safety factor $k_{\sigma y}$/1/ | **17.8** | **14.3** |

$$k_{\sigma y} = \frac{\sigma_{redMAX}}{\sigma_y} \qquad (3)$$

Note, if $k_{\sigma y} > 1$, it means safe situation (i.e. plastic deformations do not occurs). Application of safety factor is the typical way of reliability assessment in the branch of engineering. The calculated results are presented in Tables 2 and 3.

# 3   Conclusion

Distal radius fractures are the most frequent type of injury in the upper limbs. Therefore, it is good idea to focus on them. Hence, this conference paper analyses a new locking compression (LCP) plate for osteosynthesis. The plate is made of Ti6Al4V or stainless steel. The bone (with a fracture of 23–C1 AO classification), together with the plate and angularly stable screws, was exposed to axial tension/compression and bending quasi-dynamic overloading.

The material properties of bone can be described with sufficient accuracy for individual bone parts (substantia compacta and substantia spongiosa tissues) using a homogeneous isotropic material model (based on real geometry, CT, Mimics software).

Variant calculations (i.e. numerical simulations via Finite Element Method—Ansys Workbench software) were carried out for fused bone (i.e. successful treatment) and non-fused bone (i.e. unsuccessful treatment). The safety factor was evaluated with regard to the minimum yield strength of the material. The situation of maximum loading on non-fused bone is an extreme state in which the fracture fails to heal and the patient places excessive stress on the limb.

In cases of successful treatment, the safety factor $k_{\sigma y} > 14.3$. In cases of unsuccessful treatment, the safety factor $k_{\sigma y} > 7.2$ (i.e. plastic deformation do not occurs). From this biomechanical perspective, the analysed new design of locking compression plate with its angularly stable screws can be considered safe for the treatment of distal radius fractures. This is connected with improvement of medical treatment.

Future developments will involve improvements in the description of bone material properties, for example see Fig. 7, the use of a plastic model in FEM

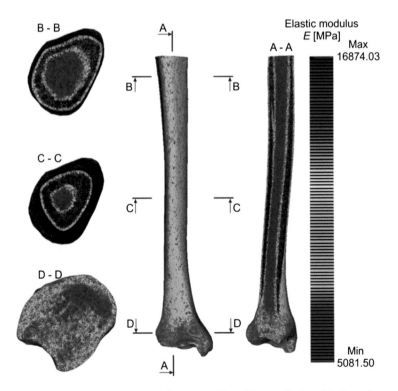

**Fig. 7** Size of elasticity modulus (100 defined material models, acquired via Mimics software and MSC.PATRAN software)—tibia

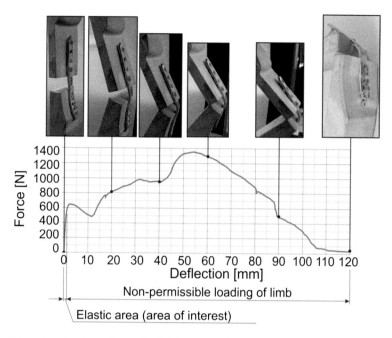

**Fig. 8** Experiments—dependence of axial force on axial deflection—tibia, medial plate

analysis, and the cyclical loading of the plate to determine limit fatigue strength. Other possibilities include the application of probability methods and experimentation on artificial bone, different plates and screws for plates etc., see Fig. 8 and Refs. [7–10].

**Acknowledgments** This work was supported by the Czech projects TA03010804 and SP2015/180.

# References

1. Frydrýšek, K., Jořenek, J., UČEŇ, O., Kubín, T., Žilka, L., Pleva, L.: Design of external fixators used in traumatology and orthopaedics—treatment of fractures of pelvis and its acetabulum. Procedia Eng. **48**, 164-173 (2012). doi:10.1016/j.proeng.2012.09.501
2. Frydrýšek, K.: Probabilistic approaches used in the solution of design for biomechanics and mining. In: Advances in Safety, Reliability and Risk Management—Proceedings of the European Safety and Reliability Conference, ESREL 2011. 2012, 1900–1905. ISBN 9780415683791
3. Frydrýšek, K., Učeň, O., Fojtík, F., Poruba, Z., Theisz, G., Sejda, F., Michenková, Š.: Numerical and experimental verifications of external and internal fixators for traumatology and orthopaedics. In: Bachorz, P., Jureczko, M., Jureczko, P. (eds.) Applied Mechanics 2014: 16th International Scientific Conference. Conference Proceedings. Wydawnictwo Katedry Mechaniki Teoretycznej i Stosowanej, Gliwice, Kraków, Poland, 14–16 Apr 2014. ISBN 978-83-60102-67-1

4. Hneider, R., Faust, G., Hindenlang, U., Helwig, P.: Review: inhomogeneous, orthotropic material model for the cortical structure of long bones modelled on the basis of clinical CT or density data. Comput. Methods Appl. Mech. Eng. **198**(27), 2167–2174 (2009). doi:10.1016/j.cma.2009.02.010
5. Helgason, B. et al.: Mathematical relationships between bone density and mechanical properties. Clin. Biomech. 135–146 (2008). 10.1016/j.clinbiomech.2007.08.024
6. Ruber, V.: Algoritmus ošetření zlomenin distálního radia s důrazem na nitrokloubní zlomeniny, doctoral dissertation (thesis supervisor: Doc. MUDr. Michal Mašek), Lékařská fakulta, pp. 1–132. Masarykova Universita, Brno (2009)
7. Theisz, G., Frydrýšek, K., Fojtík, F., Kubín, T., Pečenka, L., Demel, J., Madeja, R., Sadílek, M., Kratochvíl, J., Pleva, L.: Medial plate for treatment of distal tibia fractures. In: Experimental Stress Analysis 2015, pp. 431–437. CTU in Prague, Prague (2015). ISBN 978-80-01-05735-3
8. Vilimek, M., Sedláček, R.: Methodology for dynamic testing of plates for proximal tibia osteosynthesis. Bull. Appl. Mech. **7**(25), 12–14 (2011). ISSN 18011217
9. Losertová, M., Drápala, J., Konečná, K., Pleva, L.: Study of fracture feature of Titanium based alloys for biocompatible implants after removal from human body. Mater. Sci. Forum **782**, 449–452 (2014). ISBN 978-303835072-9
10. Frydrýšek, K., Čada, R.: Probabilistic reliability assessment of femoral screws intended for treatment of "collum femoris" fractures. In: International Conference of the Polish Society of Biomechanics, BIOMECHANICS 2014. Łódż, Poland (2014)

# The Enlarged d-q Model of Induction Motor with the Iron Loss and Saturation Effect of Magnetizing and Leakage Inductance

Jan Otýpka, Petr Orság, Vítězslav Stýskala, Dmitrii Kolosov, Stanislav Kocman and Feodor Vainstein

**Abstract** The main aim of this paper is the presentation of induction motor model with the iron loss and the saturation effect of magnetizing and leakage inductances. The previous published standard models of induction motor using the d-q model actually neglect the iron loss effect. Hence, the iron loss represents about 3–5 % total loss in the induction motor. This paper is focused on the inferred model, which considers the iron loss in motor, and comparing with the standard models. The second part of the paper deals with the saturation effect of magnetizing and leakage inductances.

**Keywords** Induction motor · D-q coordinate frame model · Iron loss · Saturation effect · Magnetizing inductance · Leakage inductance · T-form

## 1  Introduction

The conventional model of induction motor is based on the classical T-form layout without the iron loss. This model is widely used for the operation of induction motor. The conventional model is presented many researchers [1–4]. In case of the operation of such motor with minimal losses or the power dynamic analysis of induction motors powered by the harmonic or nonharmonic source, the conven-

J. Otýpka · P. Orság · V. Stýskala · D. Kolosov (✉) · S. Kocman
Department of Electrical Engineering FEECS, VŠB—Technical University of Ostrava, 17. Listopadu 15, 708 33 Ostrava, Poruba, Czech Republic
e-mail: dmitrii.kolosov@vsb.cz

J. Otýpka
e-mail: jan.otypka@vsb.cz

F. Vainstein
STEM College, Texas A&M University Texarkana, 7101 University Ave, Texarkana, TX 75503, USA
e-mail: Feodor.Vainstein@tamut.edu

© Springer International Publishing Switzerland 2016
V. Styskala et al. (eds.), *Intelligent Systems for Computer Modelling*,
Advances in Intelligent Systems and Computing 423,
DOI 10.1007/978-3-319-27644-1_2

11

tional model is not sufficient for the using. Hence, this model must be enlarged with the consideration of the iron loss.

This paper presents the enlarged model with the iron loss in d-q coordinate frame system. New model is compared with the conventional model for the starting of induction motor without a load. Each kind of loss is calculated by this model as the conventional model considers only the stator and rotor windings losses. On the other hand, the iron loss is not separated from the eddy-current loss and hysteresis loss, because the physical interpretation of both losses is relatively complicated. The possible solving of this problem is published in [5]. Therefore, the enlarged model operates only with one value of the resistance $R_{Fe}$ represented the iron loss, which includes both losses.

This paper is also focused on the saturation effect of magnetizing and leakage inductances, which are considered to be constant according to the stator and rotor currents for the standard solving. These nonlinear parameters are obtained by the identification method from the short circuit test and the idle test, which were measured on the specific type of induction motor. The model involves the blocks for direct calculation of the actual values of saturated magnetizing inductance and stator and rotor leakage inductances, which are given by nonlinear dependence on the magnetizing, stator and rotor currents. The topic conclusion is devoted to the simulation results of induction motor starts with the linear and nonlinear models.

## 2   The Model of Induction Motor with Iron Loss

The conventional model of induction motor is based on the classical equivalent circuit in T-form, from which derives the d-q coordinate frame model, presented in [1–4]. If this model is enlarged by the additional iron loss resistance $R_{Fe}$, the currents flowing in the transversal branch must be defined. The transversal branch is presented by the parallel combination of iron loss resistance $R_{Fe}$ and magnetizing inductance $L_m$. The model of induction motor with the considering of iron loos is obtained, when the model derivation is based on the magnetizing current $i_m$, which flows through the magnetizing inductance $L_m$.

The derivation of the induction motor model is obtained from Fig. 1, where the currents (the main current $i_0$, which flows through the transversal branch and the current in iron $i_{Fe}$) are defined by using the current divider. These currents are obtained as the function of the magnetizing current $i_m$. This approach for derivation of the induction motor model with iron loss is presented in [3, 6, 7].

The model of induction motor with the iron loss is given by these equations presented in [3, 6, 7] for the general d-q coordinate frame system, which rotates with the general angle speed $\omega$:

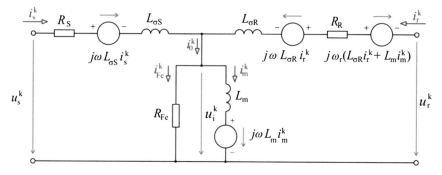

**Fig. 1** The equivalent circuit model of induction motor in T-form with the iron loss

$$u_{ds}^k = R_S i_{ds}^k + L_{\sigma S} \frac{di_{ds}^k}{dt} + L_m \frac{di_{dm}^k}{dt} - \omega\left(L_{\sigma S} i_{qs}^k + L_m i_{qm}^k\right), \tag{1}$$

$$u_{qs}^k = R_S i_{qs}^k + L_{\sigma S} \frac{di_{qs}^k}{dt} + L_m \frac{di_{qm}^k}{dt} + \omega\left(L_{\sigma S} i_{ds}^k + L_m i_{dm}^k\right), \tag{2}$$

$$u_{dr}^k = R_R i_{dr}^k + L_{\sigma R} \frac{di_{dr}^k}{dt} + L_m \frac{di_{dm}^k}{dt} - (\omega - \omega_r)\left(L_{\sigma R} i_{qr}^k + L_m i_{qm}^k\right), \tag{3}$$

$$u_{qr}^k = R_R i_{qr}^k + L_{\sigma R} \frac{di_{qr}^k}{dt} + L_m \frac{di_{qm}^k}{dt} + (\omega - \omega_r)\left(L_{\sigma R} i_{dr}^k + L_m i_{dm}^k\right), \tag{4}$$

$$i_{d0}^k = i_{ds}^k + i_{dr}^k, \tag{5}$$

$$i_{q0}^k = i_{qs}^k + i_{qr}^k, \tag{6}$$

$$i_{d0}^k = i_{dFe}^k + i_{dm}^k = i_{dm}^k + \frac{L_m}{R_{Fe}}\left(\frac{di_{dm}^k}{dt} - \omega i_{dm}^k\right), \tag{7}$$

$$i_{q0}^k = i_{qFe}^k + i_{qm}^k = i_{qm}^k + \frac{L_m}{R_{Fe}}\left(\frac{di_{qm}^k}{dt} + \omega i_{qm}^k\right), \tag{8}$$

where $\omega_r$ is the electrical rotor angle frequency. The magnetic fluxes components are given by the equations in axis d and q:

$$\Psi_{ds}^k = L_{\sigma S} i_{ds}^k + L_m i_{dm}^k, \tag{9}$$

$$\Psi_{qs}^k = L_{\sigma S} i_{qs}^k + L_m i_{qm}^k, \tag{10}$$

$$\Psi_{dr}^k = L_{\sigma R} i_{dr}^k + L_m i_{dm}^k, \tag{11}$$

$$\Psi_{qr}^k = L_{\sigma R} i_{qr}^k + L_m i_{qm}^k. \tag{12}$$

The inducted voltage on the magnetizing branch is defined by the following equations in d-q coordinate frame:

$$u_{di}^k = L_m \frac{d i_{dm}^k}{dt} - \omega L_m i_{dm}^k, \tag{13}$$

$$u_{qi}^k = L_m \frac{d i_{qm}^k}{dt} + \omega L_m i_{qm}^k. \tag{14}$$

The electromagnetic torque in general d-q coordinate frame is given by the interaction between rotor magnetic fluxes components and stator currents components:

$$T_{em} = \frac{3}{2} p_p \frac{L_m}{L_{\sigma R} + L_m} \left[ \Psi_{qr}^k \left( i_{ds}^k - i_{dFe}^k \right) - \Psi_{dr}^k \left( i_{qs}^k - i_{qFe}^k \right) \right], \tag{15}$$

The mechanical equation is given by:

$$J \frac{d\Omega_m}{dt} = T_{em} - T_L - k_F \Omega_m, \tag{16}$$

where $\Omega_m$ is the mechanical angle speed, $J$ is inertia, $T_L$ is load torque and $k_F$ is friction coefficient. The calculation of total losses is presented in [8], thereafter the total losses in model could be separated to dissipated power by the joule effect in stator and rotor winding and the iron loss. The total losses are given by:

$$\Delta P_{celk.} = \Delta P_{j.s} + \Delta P_{j.r} + \Delta P_{Fe}$$
$$= \frac{3}{2} \left[ R_S \left( i_{ds}^2 + i_{qs}^2 \right) + R_R \left( i_{dr}^2 + i_{qr}^2 \right) + \frac{1}{R_{Fe}} \left( u_{di}^2 + u_{qi}^2 \right) \right]. \tag{17}$$

The induction motor identified parameters are given in Table 1. The parameters of longitudinal branch: rotor resistance $R_R$, leakage inductances $L_{\sigma S}$ a $L_{\sigma R}$ are obtained from the short circuit test. The parameters of transversal branch: magnetizing inductance $L_m$ and the iron loss resistance $R_{Fe}$ are obtained from the idle test. These parameters are considered to be constant in simulation.

The model of induction motor with the iron loss is created in the Matlab Simulink environment. The simulation is focused on the starting process without load, when the enlarged model of induction motor with the iron loss is compared with the conventional model. Both models are simulated in stationary d-q frame for angle speed $\omega = 0$ rad/s and supply voltage has amplitude 325 V with frequency 50 Hz. The simulation results are presented at Figs. 2, 3, 4, 5, 6 and 7, where the

**Table 1** The table of identified parameters for specific type of induction motor

| Parameter | Sign | Value |
|---|---|---|
| Nominal power | $P_N$ | 4 kW |
| Nominal voltage | $U_N$ | 400 V |
| Nominal current | $I_N$ | 8.1 A |
| Nominal frequency of mains | $f_N$ | 50 Hz |
| Nominal speed | $n_n$ | 1440 rpm |
| Effectivity class | | IE2 |
| Stator resistance | $R_S$ | 1.1 Ω |
| Stator leakage inductance | $L_{\sigma S}$ | 9.5 mH |
| Rotor resistance | $R_R$ | 1.478 Ω |
| Rotor leakage inductance | $L_{\sigma R}$ | 14.8 mH |
| Magnetizing inductance | $L_m$ | 172.7 mH |
| Equivalent resistance for core loos | $R_{Fe}$ | 491 Ω |
| Inertia | $J$ | 0.02 kg m$^2$ |

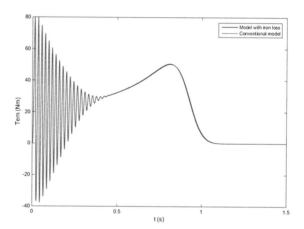

**Fig. 2** The waveform of electromagnetic torque for both models

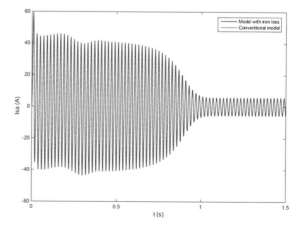

**Fig. 3** The waveform of stator current in first phase for both models

**Fig. 4** The detail of stator current waveform for both models

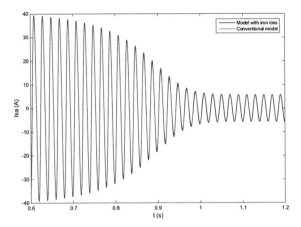

**Fig. 5** The absolute value of magnetizing current $i_m$ given by the vector sum of components in d and q axis for enlarged model of induction motor

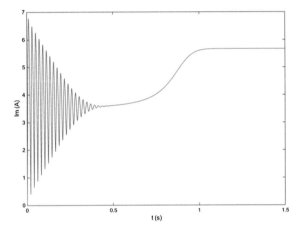

**Fig. 6** The absolute value of iron current $i_{Fe}$ given by the vector sum of components in d and q axis for enlarged model of induction motor

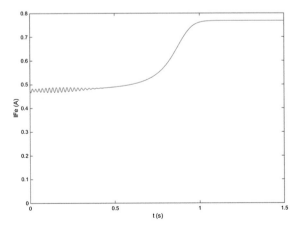

**Fig. 7** The absolute value of inducted voltage over magnetizing branch $u_i$ given by the vector sum of components in d and q axis for enlarged model of induction motor

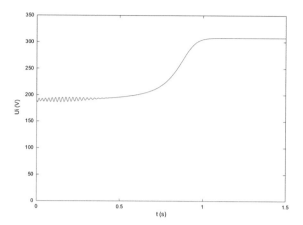

electromagnetic torque $T_{em}$, the stator current of first phase $i_{sa}$, the magnetizing current $i_m$, the iron current $i_{Fe}$ inducted voltage over magnetizing branch $u_i$ are shown.

## 3  The Saturating Effect of Magnetizing and Leakage Inductance in Induction Motor

The parameters like magnetizing inductance, stator and rotor leakage inductance are considered to be constant in the previous simulations of induction motor. Therefore, the magnetizing flux and leakage fluxes are taken as the linear functions of magnetizing current or stator and rotor currents. In fact, the magnetizing inductance is saturated, thus the magnetizing flux is nonlinear versus the magnetizing current, as it is described in [9, 10] and it is shown on Fig. 8. The similar effect could be observed for the leakage fluxes, which pass predominantly through the air. Hence, the stator and rotor leakage fluxes are conventionally expected to be linear depending on the stator and rotor currents. However, a little part of the stator and rotor leakage fluxes passes through the iron, because the leakage inductances are nonlinear to the stator and rotor current. This issue is solved in [11–13].

### 3.1  The Saturating Effect of Magnetizing Inductance

The main magnetizing flux is nonlinear to the magnetizing current of induction motor, as it is shown on Fig. 8. This fact is proved by the analysis from the idle test measured in the steady state. These nonlinear dependence of magnetizing flux $\Psi_m$ and magnetizing current $i_m$ can be approximated by the odd order polynomial function. The reason is that the supply voltage consists of the characteristic

**Fig. 8** The nonlinear dependence of magnetizing flux $\Psi_m$ on magnetizing current $i_m$

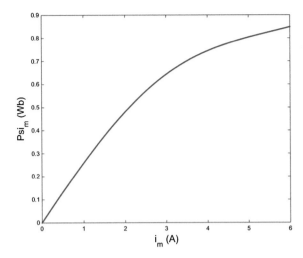

harmonic frequencies with general frequency $f_1$ and only odd harmonic frequencies $3f_1$, $5f_1$, $7f_1$, ... $pf_1$, which are generated by the source and nonlinear magnetizing branch [10]. The nonlinear dependence between magnetizing flux and current can be approximated by the following function:

$$\Psi_m = k_1 i_m + k_3 i_m^3 + k_5 i_m^5 + k_7 i_m^7 + \cdots, \qquad (18)$$

where $k_1$, $k_3$, $k_5$, etc. are coefficients of polynomial function. The magnetizing inductance is defined by form:

$$L_{m,d} = \frac{d\Psi_m}{di_m}. \qquad (19)$$

The nonlinear dependences of the magnetizing flux $\Psi_m$ and the magnetizing inductance $L_m$ on the magnetizing current $i_m$, which are obtained by the direct approximation from the measuring and parameters identification, these are shown on Figs. 8 and 9.

## 3.2 The Saturating Effect of Leakage Inductances

The nonlinear dependence is also determined for both leakage inductances $L_{\sigma S}$ and $L_{\sigma R}$. This fact was verified from the short circuit test measured with reduced voltage. It is evident in Fig. 11 that the leakage inductances have higher values in the range of low currents comparing with the nominal current. This important fact must be included to the model of induction motor. This issue is described in [11–13].

**Fig. 9** The nonlinear
dependence of magnetizing
inductance $L_m$ on magnetizing
current $i_m$

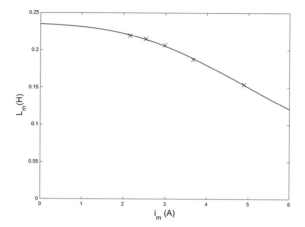

**Fig. 10** The nonlinear
dependence of leakage fluxes
on stator current

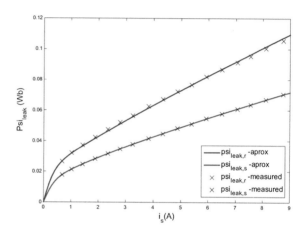

**Fig. 11** The nonlinear
dependence of leakage
inductances on stator and
rotor current

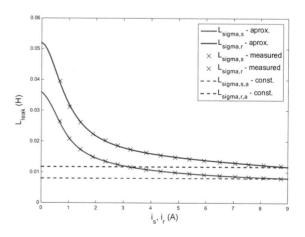

The stator and rotor leakage inductance are separated into the constant components and nonlinear components according to [11, 12]. Thereafter, the leakage inductances are given by:

$$L_{\sigma S} = L_{\sigma S,a} + L_{\sigma S,i}(i_s), L_{\sigma R} = L_{\sigma R,a} + L_{\sigma R,i}(i_r), \tag{20}$$

where $L_{\sigma S,a}$ and $L_{\sigma R,a}$ are the linear components of leakage inductances in the air and $L_{\sigma S,i}(i_s)$ and $L_{\sigma R,i}(i_r)$ are nonlinear components of leakage inductances in the iron, which depend on the stator and rotor currents. The nonlinear dependences of the leakage fluxes and inductances on currents are shown on Figs. 10 and 11, which were obtained by the short circuit test.

## 3.3   Simulation Results

The model of induction motor with the iron loss according to Eqs. (1)–(17) was enlarged by the saturating effect of magnetizing inductance $L_m$ and leakages inductances $L_{\sigma S}$ and $L_{\sigma R}$. Each inductance is implemented by the polynomial function to the model, which calculates the values of inductances from the actual values of each current. The general schema is presented on Fig. 12.

The model with the iron loss and constant parameters is compared with the both models, which includes the saturation effect of magnetizing inductance and leakage inductances (model with only saturated magnetizing inductance and model with each saturated inductance). The simulation results are better illustrated by the electromagnetic torque waveform, see Fig. 13.

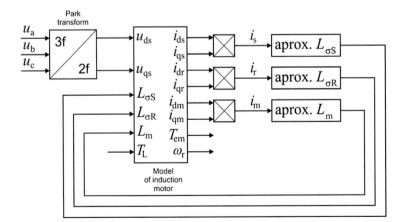

**Fig. 12** The model of induction motor with calculation of magnetizing, stator and rotor leakage inductance

**Fig. 13** The electromagnetic
torque waveform simulated
by introduced models

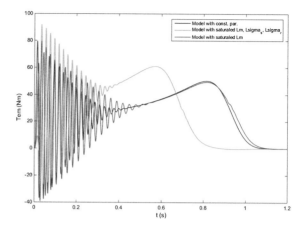

## 4 Conclusion

The simulations results show than the differences between the conventional model
and the enlarged model with iron loss are negligible for normal operating conditions. The minor differences may appear in dynamic states. This fact is shown in
Figs. 2, 3 and 4, where the starting process of induction motor is shown. The correct
solution of model with iron loss is confirmed, as the iron current is much smaller
than magnetizing current ($i_{Fe} \ll i_m$). Therefore, the iron current does not consid-
erably affect the stator current, as is shown in Figs. 3 and 4. The simulated iron
current caused the power dissipation 350 W in the steady state, confirms the values
measured during the idle test.

It is evident in Fig. 13, that the nonlinear dependence of inductances on currents
have major impact for simulation of torque, motor currents and speed waveforms,
because the electrical time constant is dynamic changed due to the inductances
calculations. The saturating effect has the major impact on the dynamic changes in
the models of induction motor against the model with constant parameters.

**Acknowledgments** The work is partially supported by Grant of SGS No. SP2015/151, VŠB—
Technical University of Ostrava, Czech Republic.

## References

1. Krause, P.C., Wasynczuk, O., Sudhoff, S.D.: Analysis of Electric Machinery and Drive
   Systems. IEEE Press, NY (2002)
2. Leedy, A.W.: Simulink/MATLAB Dynamic Induction Motor Model for use in Undergraduate
   Electric Machines and Power Electronics Courses. IEEE, NY (2013)
3. Neborák, I.: Modelování a simulace elektrických regulovaných pohonů. Monografie, VŠB-TU
   Ostrava 2002, 172 stran, ISBN 80-248-0083-7

4. Giri, F.: AC Electric Motors Control: Advanced Design Techniques and Applications. Wiley (2013) ISBN 978-1-118-33152-1
5. Kioskeridis, I., Margaris, N.: Loss minimization in induction motor adjustable-speed drives. IEEE Trans. Ind. Electron. **43**, 226–231 (1996)
6. Jung, J.: A vector control scheme for EV induction motors with a series iron loss model. IEEE Trans. Ind. Electron. **45**, 617–624 (1998)
7. Moulahoum, S., Touhami, O.: A Satured Induction Machine Model with Series Iron Losses Resistance. IEEE, Power Engineering, Energy and Electrical Drives (2007)
8. Lim, S., Nam, K.: Loss-minimising control scheme for induction motors. IEE Proc. Electr. Power Appl. **151**, 385–397 (2004)
9. Levi, E.: A unified approach to main flux saturation modeling in D-Q axis models of induction machines. IEEE **10**, 455–461 (1995)
10. Kerkman, R.J.: Steady-state and transient analysis of an induction machine with saturation of the magnetizing branch. IEEE Trans. Ind. Appl. **21**(1), 226–234 (1985)
11. Lipo, T.A., Consoli, A.: Modeling and simulation of induction motors with saturable leakage reactances. IEEE fians. Ind. Appl. **IA-20**(1), 180–189 (1981)
12. Keyhani, A., Tsai, H.: IGSPICE simulation of induction machines with saturation inductance. IEEE Trans. Energy Convers. **4**(1) (1989)
13. Alsammak, A.N.B., Thanoon, M.F.: An improved transient model of an induction motor including magnetizing and leakage inductances saturated effect. Int. J. Eng. Innovative Technol.**3**(10) (2014)

# Fuzzy Model Based Intelligent Prediction of Objective Events

Sergey Kovalev, Andrey Sukhanov and Vitêzslav Stýskala

**Abstract** There are many processes under control. Nowadays all of them undergo through the automation procedure due to the rapid development of information technologies and computing devices. Because of the fact that many tasks of control deal with the elimination of situations, which can cause emergencies, automatic prediction of objective events in process streams becomes the popular problem decided by automation procedure. This work presents the new approach for intelligent analysis of processes represented by time series data. The main aim of our approach is prediction and detection of objective events. The idea of the technique is referred to the mapping of original time series into phase space and construction of the fuzzy prediction clusters of patterns in this phase space. Prediction stage consists of comparison of observed event with the prediction clusters according to the base of fuzzy rules.

**Keywords** Time-series prediction models · Automation · Fuzzy clusters · Temporal mapping

## 1 Introduction

Time series data analysis is the key task of many engineering and economical tasks [1–3]. One of the most important aims of time series analysis is prediction of interesting temporal structures, which are called as temporal patterns. Temporal patterns often belong to the events, which are objective in the definition of a certain problem formulation. Objective events represent particular values or value intervals from observed time series, which allow us to make decision about process. Particularly, objective events can be referred to outliers, which are rather different

S. Kovalev · A. Sukhanov
Rostov State Transport University, Rostov-on-Don, Russia
e-mail: drewnia@rambler.ru

V. Stýskala (✉)
VŠB Technical University of Ostrava, Ostrava, Czech Republic
e-mail: vitezslav.styskala@vsb.cz

© Springer International Publishing Switzerland 2016
V. Styskala et al. (eds.), *Intelligent Systems for Computer Modelling*,
Advances in Intelligent Systems and Computing 423,
DOI 10.1007/978-3-319-27644-1_3

from the typical patterns representing the normal behavior. In practical point of view, the topic of time series analysis for detection of objective events is related to the technological process control problems, such as prediction of emergencies, diagnostics of the pre-fault states of technical devices, forecasting of telecommunication traffic, intrusion detection in computer network, etc. The popular way for decision of above mentioned problem is intelligent analysis of time series (or Intelligent Time Series Data Mining). This procedure implies the discovering useful information from time-series data, which is hidden and unknown in normal circumstances. It combines tools from different areas, including machine learning, statistics and database design, and applies many techniques, such as clustering, association rules, visualization, etc. [4, 5]. Intelligent analysis is one of the popular approach for processing of poorly structured objects, which have the unsteadiness, nonlinearity of analyzed processes and their "disobedience" to typical distribution laws as well. In this case, additional empirical information, which is obtained from human experts, and using of the artificial intelligence methods for processing of this information are required.

Present paper proposes the new approach for intelligent analysis of time series, which is based on usage of fuzzy logic for modeling of human experts empirical knowledge about the process under control. The main aim of the proposed approach is prediction of objective events. Key idea of the method was inspired by the reconstructed phase space framework for event prediction [3, 6, 7]. The paper is organized as follows. Section 2 presents the notion of characterization function, which is used for objective events representation. Section 3 provides the idea of importance of the mapping of time series into phase space for the processing. The function for predictive cluster construction is described in Sect. 4. We introduce the fuzzy model for objective events prediction in Sect. 5. Section 6 describes the example of implementation for our technique. Conclusions and future work are discussed in Sect. 7.

## 2 Characterization Functions of Objective Events in Time Series

Time series can be presented as a sequence of time steps:

$$Z = \{x_t \mid t = 1, 2, \ldots, N\}, \tag{1}$$

where $t$ is the time index and $N$ is number of observations.

Objective event of time series can be described as any given event $x_t$, which is pre-labeled as objective or has the pre-defined specific feature. Objective events may have different interpretation, which depends on application. For example, the objective event can be presented by peaks of the seismic activity of the Earth's interior, price jumps on stock, limit values of current or voltage, maximal delays in train schedule, etc. Establishment of characterization function $g(x_i)$ is used for prediction of the objective events in time series. The form of characterization function depends on application, nature of time series and prediction aims. In common case, characterization function is determined by the fact that the event at current time $t$ should be correlated with the event at the certain time in future [1]. For example, if we need predict the peak value of current at the next time step, then the characterization function is expressed as

$$g(x_t) = x_{t+1}. \tag{2}$$

Obviously, if we would like to predict the peaks, we should label all $x_t$, for which $g(x_t) > c$, $c$ is constant value.

# 3 Time Series Representation in Phase Space

Intelligent Data Mining of Time Series is the developing discipline, which uses the set of different techniques to reach such aims as revealing of hidden regularities in data, anomaly detection and prediction. One of the most advanced techniques is based on both the reconstruction of dynamic system, which generates initial set of events, and future construction of model for recognition and prediction. Reconstructed presentation is obtained by mapping of initial series into $m$-dimensional metric space with using of time delay embedding. Embedding dimension $m$ and time delay $\tau$ are the main parameters of reconstructed phase space. Takens [6] shows that the correct choice of these parameters gives us phase space, which is homeomorphic to the initial time series. At the same time, recovered dynamics of system is identical to the initial dynamics and the outcomes are right. Therefore, dynamic prediction of time series can be performed by using the attractor of recovered dynamic system. Idea of transformation of time series into phase space for objective events prediction was described by many works [1, 3, 8]. In particular, the authors in [1, 3] described different problems in area of analysis of seismic and econometric time series, which are related with prediction and stock price prediction. The author used techniques, which are based on series mapping into phase space, where prediction clusters of temporal patterns are made. Optimal cluster with radius $r$ and center $c$ can be find by using genetic algorithm optimization of the objective function $F(r, c)$. Despite to novelty of such idea, this approach has the set of shortcomings. Specifically, prediction model is not robust and tightly related with structure of time series because of fixed cluster parameters. For elimination of such shortcoming, the authors in [7] presented the new approach for data mining of time series, which is based on using of fuzzy prediction clusters in phase space with using of Gaussian membership functions. Optimization algorithm based on gradient descent is used for searching of optimal parameters of clusters. In [9], authors also introduced fuzzy logic tools for reconstruction of phase space and increasing of robustness. However, the method has one significant shortcoming, which is concluded in missing guidelines about the construction of aim functions and fuzzy function organization. As a result, using rules do not take into account a set of important relations between parameters of prediction clusters and information about time series, which actually decreases the general effectiveness of the algorithms and make the quantitative estimation be impossible. Proposed technique for objective events prediction in numerical time series is also based on the mapping idea for phase attractor of generating system with the using of fuzzy classification model for adaptation properties improving and robustness increasing. The original objective function is proposed for optimization of cluster parameters. In our case, this function sets the direct relation degrees between cluster parameters and probabilities of objective events, which is useful for quantitative estimations. Prediction is performed by fuzzy inference, which permits to use the following clarifications based on additional statistical or another analysis for prediction. In this case, the fuzzy-stochastic system for prediction can be made.

Phase space is organized by patterns $X_t = \left(x_{t-(m-1)\tau}, x_{t-(m-2)\tau}, \ldots x_t\right)$, which are extracted from original time series in form of subsequences. These subsequences contain certain number of $m$ steps, which are divided by delay of length $\tau$ from each other. The phase space $\mathbb{R}^m$ can be presented as follows:

$$
X = \begin{pmatrix}
x_{1+(m-1)\tau} & \cdots & x_{1+\tau} & x_1 \\
x_{2+(m-1)\tau} & \cdots & x_{2+\tau} & x_2 \\
\vdots & \vdots & \vdots & \vdots \\
x_N & \cdots & x_{N-(m-2)\tau} & x_{N-(m-1)\tau}
\end{pmatrix},
\tag{3}
$$

Each row presented by pattern $X_{t+(m-1)\tau}^m = \left(x_t, x_{t+\tau}, \ldots, x_{t+(m-1)\tau}\right) \in \mathbb{R}^m$, ($t = 1, 2, \ldots, N - (m-1)\tau$), which is represented by the point in phase space $\mathbb{R}^m$.

The mutual information [10, 11] for searching the optimal $\tau$ is computed by:

$$
M(x_t, x_{t+\tau}) = \sum_{i,j} p_{ij}(\tau) \ln \frac{p_{ij}(\tau)}{p_i p_j},
\tag{4}
$$

where $x_t + \tau$ and $x_t$ are the points in the original state space obtained with delay of $\tau$, $p_i$ is the probability that a point $x_t$ is in the $i$th interval and $p_{ij}(\tau)$ is the joint probability that $x_t$ falls into the $i$th interval at time $t$ and the $j$th interval at the time $t + \tau$. The mutual information method considers the first minimum of (4) as the optimal time delay $\tau$.

The technique for searching of the optimal $m$ is called as false nearest neighbor [12]. Here, for each data point denoted by $X_i^m$ in $\mathbb{R}^m$, its nearest neighbor $X_j^m$ can be found by

$$
X_j^m = \arg \max_{X_j^m, i \neq j} \|X_i^m - X_j^m\|,
\tag{5}
$$

where $\|X_i^m - X_j^m\|$ is the Euclidean distance between the two points. When the embedding dimension is increased from $m$ to $m+1$, the changing rate of distance between the two points is calculated by:

$$r_i = \sqrt{\frac{\|X_i^{m+1} - X_j^{m+1}\|^2 - \|X_i^m - X_j^m\|^2}{\|X_i^m - X_j^m\|^2}}. \tag{6}$$

If $r$ exceeds a given threshold $p$, then $X_i^m$ is marked as having a false nearest neighbor. So, the criterion for adequate embedding dimension $m$ is determined by the fact that the number of data points, for which $r_i > p$, is zero in $\mathbb{R}^m$.

## 4 Objective Function Construction

Precursors of the objective events can be represented by temporal patterns $X_t^m \in \mathbb{R}^m$, which are labeled by appropriate values of characterization function $g(x_t)$. Here and hereafter we call precursors as predictive patterns in phase space. Collection of precursors in phase space forms the predictive clusters. Unfortunately, predictive clusters, which contain only predictive patterns without any contaminants, are not realizable because of unsteadiness, noise presence and data inaccuracies. Therefore, the optimization task for searching of approximately optimal predictive clusters appears. Criterion of optimality can be represented by the objective function, which depends on numerical cluster parameters. Commonly, these parameters are the cluster center and its size (or radius).

Specific form of objective function depends on current problem details and substantially influence on effectiveness of decision. In [3], authors describe the typical three objective functions for different problems, which are connected with prediction and detection of patterns in time series data. However, these functions are usable only for prediction of future values and cannot provide the ability of observing the probability or another quantitative estimation. The function for observing of quantitative estimations is described below. Let the characterization function be in binary form:

$$g(x_t) = \begin{cases} 1, & \text{if } X_t^m \in \mathbb{R}^m \text{ is precursor} \\ 0, & \text{if otherwise} \end{cases}, \tag{7}$$

Let $F$ be a cluster of analyzing time series, which has the center $c \in \mathbb{R}^m$ and the radius $r \in \mathbb{R}\ominus$. Let $S_+^F$ be a set of prediction patterns in the cluster $F$, and $S_\pm^F$ be a set of all patterns in the cluster. Then, if we consider the uniform distribution of patterns in the cluster $F$, the value of $P(g(x_t) = 1 | X_t \in F) = \frac{S_+^F}{S_\pm^F}$ can characterize the probability that pattern in this cluster is predictive. The optimization of such function leads to radius decreasing, and hence, to decreasing of ability to identify as more predictive patterns as possible. Because of this, it is needed to incorporate the value of $P(X_t \in F|g(x_t) = 1) = \frac{S^F_+}{S^{\mathbb{R}^m}_+}$ because it characterizes the percentage of those predictive patterns, which lie to the cluster. In other words, the value of $P(X_t \in F|g(X_t) = 1)$ shows the probability that predictive pattern falls into the observed cluster. Therefore, the value $P(g(x_t) = 1) = P(g(X_t) = 1|X_t \in F) \cdot P(X_t \in F|g(x_t) = 1)$ characterizes the probability of identification of random predictive pattern in phase space and can be used as objective function. Resulting expression for objective function is presented as follows

$$F(c,r) = \frac{|S^F_+|}{|S^F_{\pm}|} \cdot \frac{|S^F_+|}{|S^{\mathbb{R}^m}_+|}, \tag{8}$$

where $S^F_+ = \{X^m_t \in \mathbb{R}^m | g(x_t) = 1, \|X^m_t - c\| < r\}$, $S^F_{\pm} = \{X^m_t \in \mathbb{R}^m | \|X^m_t - c\| < r\}$ and $S^{\mathbb{R}^m}_+ = \{X^m_t \in \mathbb{R}^m\}$.

Let us describe the situation, when the cluster $F$ of phase space is the fuzzy subset of points in form of $m$-dimensional sphere with fuzzy parameters of both center $C$ and radius $R$, which are characterized by membership functions $\mu_C(X)$ and $\mu_R(X^m_t - X^m)$.

Membership function, which expresses that current point of phase space $X^m_t$ falls into a fuzzy cluster $F$, can be computed as

$$\mu_F(X^m_t) = \sup_{X^m_i \in \mathbb{R}^m} \mu_C(X^m_i) \mathbin{\&} \mu_R(\|X^m_t - X^m_i\|), \tag{9}$$

where $\mu_C(X^m_i)$ is the membership function of point $X^m_i$ in the fuzzy set representing the center $C$ of the cluster $F$, $\mu_R(\|X^m_t - X^m_i\|)$ is the membership function of distance in the fuzzy set representing the radius $R$ of the cluster $F$ and operator "&" is a fuzzy conjunction.

Such representation not only allows us to take into account ambiguous of data, when clusters are formed, but also provides the ability to use the qualitative information collected from human experts.

The number of patterns, which fall into fuzzy cluster $F$, is defined by the fuzzy estimation of capacity $|\widetilde{S}^F_{\pm}| = \sum_{X^m_t \in \mathbb{R}^m} \mu_F(X^m_t)$. The number of predictive patterns, which fall into fuzzy cluster $F$, can be defined by the fuzzy variable $|\widetilde{S}^F_+| = \sum_{X^m_t \in \mathbb{R}^m} \mu_F(X^m_t) \cdot g(x_t)$. The number of all predictive patterns in phase space is defined as $|\widetilde{S}^{\mathbb{R}^m}_+| = \sum_{X^m_t \in \mathbb{R}^m} g(x_t)$. As a result, we can summarize (8) for the case of fuzzy representation of predictive clusters. Resulting equation for fuzzy objective function expressed by:

$$F(C,R) = \frac{|\widetilde{S}^F_+|}{|\widetilde{S}^F_{\pm}|} \cdot \frac{|\widetilde{S}^F_+|}{|\widetilde{S}^{\mathbb{R}^m}_+|} = \frac{\left(\sum_{X^m_t} \mu_F(X^m_t) \cdot g(x_t)\right)^2}{\sum_{X^m_t} \mu_F(X^m_t) \cdot \sum_{X^m_t} g(x_t)}. \tag{10}$$

Note that objective function (10) always has the less values than in case of crisp clusters (8), because crisp estimation in (10) is multiplied by membership function values.

## 5 Fuzzy Model for Objective Events Prediction

Prediction of proposed algorithm is performed by comparison of new temporal patterns, which are on the input of model, to the fuzzy predictive clusters in reconstructed phase space.

Key object of the prediction system is fuzzy classification model, which is based on fuzzy rules. The fuzzy rules describe the comparison logic of current temporal pattern with predictive clusters. In the rules, fuzzy relations between parameters of temporal patterns and fuzzy clusters are antecedents and fuzzy probabilistic esti-mations of objective events predictions are consequents. Decisions about poten-tiality of future objective events are made according to fuzzy inference laws. Figure 1 depicts the structure of the prediction system.

The main principle of the model is existence of two types of rules. The first rules group consists of universal rules, which describe the obvious dependencies between predictive clusters and temporal patterns when objective events are coming. For example, it is obvious that increasing the cluster radius $R$ and decreasing the Euclid distance $V$ from pattern to the center of cluster lead to increasing of probability $P$ that observed pattern is precursor. Based on this assumption, the next universal rule for classification model can be formulated:

$$\text{IF } R \text{ is BIG AND } V \text{ is SMALL THEN } P \text{ is BIG,} \tag{11}$$

where BIG and SMALL are fuzzy variables with membership functions, which are established using the expert knowledges.

**Fig. 1** The general structure of objective event prediction system

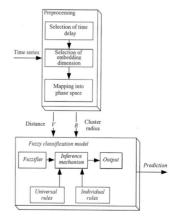

This group also consists of Takagi-Sugeno fuzzy rules, which allow to assume the objective function values. Sample rule can be constructed as

$$\text{IF } R \text{ is BIG AND } V \text{ is SMALL THEN } P_{max} = \frac{\sum_{X_t^m} \mu_F\left(X_t^m\right) \cdot g(x_t)}{\sum_{X_t^m} \mu_F\left(X_t^m\right)}. \tag{12}$$

The second rules group is based on individual fuzzy rules, which depend on certain application. Individual rules are made according to expert knowledge, which is based on qualitative notions about the specifications of time series. Antecedents of individual rules can be based on particular values of temporal patterns. As a sample, the next assumption can express the individual fuzzy rule:

$$\text{IF } V \text{ is BIG AND } x_t \text{ is POSITIVE SMALL THEN } P \text{ is BIG,} \tag{13}$$

where $x_t$ is the current value of time series.

## 6  Implementation Example

As an example of implementation, the prediction for time series generated by the Rössler system can be used [13]. The Rössler system is defined by three non-linear ordinary differential equations, which define a continuous-time dynamical system that exhibits chaotic dynamics associated with the fractal properties of the attractor. These equations are represented as follows:

$$S(X) = \begin{cases} \frac{dx}{dt} = -y - z \\ \frac{dy}{dt} = x + ay \\ \frac{dz}{dt} = b + z(x - c) \end{cases}, \tag{14}$$

where $x$, $y$ and $z$ are the variables and $a$, $b$ and $c$ are the properties.

We examined this system for parameter values $a = 0.15$, $b = 0.2$ and $c = 10$ (similarly as was in [11]) with using MATLAB toolboxes. The trajectory was generated from an initial condition of $(-0.7, 0.7, 1)$. Figure 2 depicts the part of the Rössler time series.

To check the efficiency of the approach, the task was established as one of noticed [see Eq. (2)] in Sect. 2. As a results of preprocessing, the optimal parameters of phase space $m = 4$ and $\tau = 36$ were determined by using (6) and (4), respectively. To make the optimal clusters Eq. (10) was used as aim function. The rule base was formed by using the obvious notions about relations between patterns and clusters when objective events are coming. Moreover, we added one rule based on the individual behavior of the Rössler system. Key properties for prediction were parameters of clusters $R$ and $C$, Euclid distance from current pattern to the center of cluster $V$ and current value $x_t$. All of them were fuzzified and then fuzzy linguistic

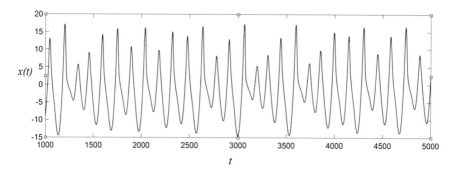

**Fig. 2** Fragment of the time series generated by the Rössler system

**Fig. 3** The table of Sugeno rules

| R \ V | S | M | B |
|---|---|---|---|
| S | $0.7\,P_{max}$ | $0.9\,P_{max}$ | $P_{max}$ |
| M | $0.3\,P_{max}$ | $0.7\,P_{max}$ | $0.9\,P_{max}$ |
| B | $0.1\,P_{max}$ | $0.3\,P_{max}$ | $0.7\,P_{max}$ |

$$P_{max} = \frac{\sum_{X_t^m} \mu_F(X_t^m) \cdot g(x_t)}{\sum_{X_t^m} \mu_F(X_t^m)}$$

values were used as input parameters for the model. We used three linguistic variables (L—low, M—medium, B—big) for $R$ and $V$, which are defined on the interval of $[0, 60]$, and five linguistic variables (NB—negative big, NS—negative small, Z—zero, PS—positive small, PB—positive big) for $x_t$, which are defined on the interval of $[-15, +20]$. All of the linguistic variables were defined in form of triangular membership functions.

Sugeno fuzzy rules included into model are illustrated by Fig. 3. Rows and columns are the antecedents and intersections are the consequents.

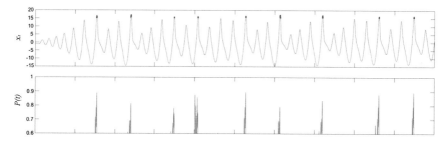

**Fig. 4** The part of prediction results

The individual fuzzy rule was defined as:

$$\text{IF } V \text{ is B AND } x_t \text{ is PB THEN } P = P_{max}. \tag{15}$$

Figure 4 depicts the prediction results, which are made by using the fuzzy rules base.

## 7 Conclusions and Future Work

Presented approach for intelligent data mining, which is aimed to predict objective events, has more advantages than previous ones because of some reasons. At first, it is more robust for noises and more adaptive for new values because of fuzzy rules usage. At second, the qualitative expert analysis, which reflects the visual presentation of time series in phase space, can be integrated by using fuzzy clusters in the model. At third, developed objective function, which sets the direct relation between patterns and predictive clusters, provide the ability of observing of the quantitative predictive estimations and their using for the future analysis. Future research is needed to extend the system for specific application, for example, in prediction of delay in railway transport. Moreover, it can be supplemented by the probabilistic model for detailed processing of certain time series.

**Acknowledgments** This work was supported by the Russian Foundation for Basic Research (Grants No. 13-07-00183 A, 13-08-12151 ofi_m_RZHD), by SGS, VSB-Technical University of Ostrava, under the grant no. SP2014/110 and partially supported by Grant of SGS No. SP2015/151, VŠB—Technical University of Ostrava, Czech Republic.

## References

1. Povinelli, R.J.: Time series data mining: identifying temporal patterns for characterization and prediction of time series events. Doctoral dissertation, Faculty of the Graduate School, Marquette University (1999)
2. Sukhanov, A.V., Kovalev, S.M.: Anomaly detection based on Markov chain model with production rules. Programmnyye produkty i sistemy **3**, 40–44 (2014)
3. Povinelli, R.J., Feng, X.: Data mining of multiple nonstationary time series. Artif. Neural Netw. Eng. 511–516 (1999)
4. Fayyad, U.M., Piatetsky-Shapiro, G., Smyth, P., Uthursamy, R.: Advances in Knowledge Discovery and Data Mining. AAAI Press, Menlo Park, Calif (1996)
5. Weiss, S.M., Indurkhya, N.: Predictive Data Mining: A Practical Guide. Morgan Kaufmann, San Francisco (1998)
6. Takens, F.: Detecting Strange Attractors in Turbulence. Springer, Berlin, pp. 366–381 (1981)
7. Feng, X., Huang, H.: A fuzzy-set-based reconstructed phase space method for identification of temporal patterns in complex time series. Knowl. Data Eng. IEEE Trans. **17**(5), 601–613 (2005)

8. Hagan, M.T., Demuth, H.B., Beale, M.H.: Neural Network Design. Pws Pub, Boston, pp. 2–14 (1996)
9. Aydin, I., Karakose, M., Akin, E.: The prediction algorithm based on fuzzy logic using time series data mining method. World Acad. Sci. Eng. Technol. **51**(27), 91–98 (2009)
10. Kantz, H., Schreiber, T.: Nonlinear Time Series Analysis, vol. 7. Cambridge University Press, Cambridge (2004)
11. Fraser, A.M., Swinney, H.L.: Independent coordinates for strange attractors from mutual information. Phys. Rev. A **33**(2), 1134 (1986)
12. Kennel, M.B., Brown, R., Abarbanel, H.D.: Determining embedding dimension for phase-space reconstruction using a geometrical construction. Phys. Rev. A **45**(6), 3403 (1992)
13. Rössler, O.E.: An equation for continuous chaos. Phys. Lett. A **57**(5), 397–398 (1976)

# Using Multiple Scanning Devices for 3-D Modeling

Feodor Vainstein, Donald Peterson and Dmitrii Kolosov

**Abstract** Medical applications for 3D printing are expanding rapidly and are expected to revolutionize health care (Hammoudi et al. in Extracting wire-frame models of street facades from 3D point clouds and the corresponding cadastral map, Saint-Mandé, France, pp. 91–96, 2010) [1]. Medical uses for 3D printing, both actual and potential, can be organized into several broad categories, including: tissue and organ fabrication; creation of customized prosthetics, implants, and anatomical models; and pharmaceutical research regarding drug dosage forms, delivery, and discovery (Sitek in IEEE Trans Med Imag 25:1172, 2006) [2]. The application of 3D printing in medicine can provide many benefits, including: the customization and personalization of medical products, drugs, and equipment; cost-effectiveness; increased productivity; the democratization of design and manufacturing; and enhanced collaboration (Bernardini in Comput Graph Forum 21 (2):149–172, 2002) [3]. Reconstruction of the object from 3D scans can be achieved either by use of sophisticated algorithms (Ozbolat and Yu in IEEE Trans Biomed Eng 60(3):691–699, 2013) [4] or directly from the point clouds (Hoy in Med Ref Serv Q 32(1):94–99, 2013) [5], (3D Print Exchange in National Institutes of Health, 2014) [6]. The second approach has an advantage of much higher speed since no image recognition is necessary. However it may also result in the loss of accuracy. To speed-up the scanning procedure we propose use of multiple scanners to obtain a point cloud of a given object. A few mathematical problems will arise with this approach. The most important among them is the calibration of multiple

F. Vainstein (✉) · D. Peterson
STEM College, Texas A&M University Texarkana, 7101 University Ave,
Texarkana, TX 75503, USA
e-mail: Feodor.Vainstein@tamut.edu

D. Kolosov
Department of Electrical Engineering, Faculty of Electrical Engineering
and Computer Science, VŠB—Technical University of Ostrava,
Ostrava, Czech Republic
e-mail: dmitrii.kolosov@vsb.cz

© Springer International Publishing Switzerland 2016
V. Styskala et al. (eds.), *Intelligent Systems for Computer Modelling*,
Advances in Intelligent Systems and Computing 423,
DOI 10.1007/978-3-319-27644-1_4

scanners. It is considered in the paper. We propose mathematical formulation of the calibration problem and give a linear time complexity algorithm to approximately solve this problem. The other problems including the study of how the measurement errors propagate to the errors of the image and how to recalculate point clouds from different scanner, is the subject of our current research and are not considered in the paper.

**Keywords** Calibration · 3-D printing · 3-D scanners · Applied mathematics · Barycentric coordinates · Linear algebra

# 1   Introduction

Recent developments in computational power and portable 3-D scanning technology make it possible to model complex human anthropometries for use in the design of tools or novel medical devices. The Microsoft KINECT (*Xbox 360* S, *Kinect* Sensor, *Microsoft Corporation* Redmond, WA) is a low-cost system that seamlessly combines CCD vision with infrared scanning and can be programmed to operate under a number of experimental conditions via open-source protocols available from Microsoft. It is well known that the KINECT system can be used to construct 3-D virtual replicates of physical objects of various sizes; however, the scanning process is not accomplished in real time and, for accuracy, scanned objects must remain stagnant during the scanning period [7]. If multiple KINECTs are used, then special markers need to be attached to the physical object in order to yield integrative data, which can be problematic for cases where marker placements may adversely impact experimental measures.

In this paper, we propose the use of multiple KINECT units to create 3-D scans without the utilization of markers, where the focus is on the calibration of multiple units in order to achieve maximum similarity among units. Following calibration, the units can simultaneously scan an object to create highly accurate virtual models that are suitable for design, computer simulations, or 3-D printing.

The paper is organized the following way. We introduce to the problem—obtaining real time scans of 3D objects and also introduce the target hardware. However results presented in the paper are general and can be used for any sort of measurement devices.

In the second part we outline the procedure of using multiple scanner.

In the third part we give a mathematical definition of software calibration and present an algorithm for calibrating.

In the fourth part we summarize the results and references on information about medical applications for 3D printing.

## 2 Usage of Multiple Scanning Devices

3-D scanner is a measurement device (MD) that can scan a portion of object visible to this device and output a file locations of points on the surface of the object visible to the scanner in the coordinate's local to the device.

To get the coordinates of the points that are not visible from current location we have to either move the MD of use multiple MDs. The second approach is more preferable since it give the opportunity to substantially speed up the scanning process.

There are two problems related to this approach.

1. Recalculating from the local coordinates of MDi to some global coordinates.
2. MDs have to be calibrated to minimize deviations in their measurements.

The first problem is relatively easy to solve.

Let us denote by s a point on the surface of S,

$v_{01}$  a vector with location of $MD_0$ as the starting point and $MD_1$ at the ending point,

$v_1$  a vector with location of $MD_1$ as the starting point and s as the ending point,

$v_0$  a vector with location of $MD_0$ as the starting point and s as the ending point. This is a vector representation of a point that may not be visible by $MD_0$.

It is clear that $v_0 = v_{01} + v_1$ (see Fig. 1).

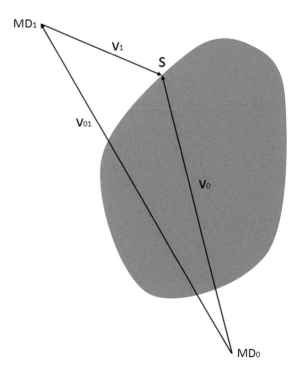

**Fig. 1** Vector diagram with the visible surface of object

## Note 1

Coordinates of the vector $v_1$ have to be recalculated to obtain co obtain coordinates in in $MD_0$ coordinate system. For this we have to know orientation of $MD_1$ relative to the position of $MD_0$. Coordinates of $v_{01}$ are given in the system of coordinates local to $MD_0$. As the result we obtain the coordinate of the point s in the coordinates local to $MD_0$.

## Note 2

If the positions of $MD_0$ and $MD_1$ are fixed, the measurements of $v_0$ and recalculation of the coordinates of $v_1$ is not difficult. However if the measurement devises are moving it will create additional problem of finding their positions and orientations relative to $MD_0$ with sufficient precision.

## Note 3

All the above is true when we use any number of MDs.

## Note 4

When a point is seen by multiple MDs, all recalculated coordinates of the point are placed in the same "master file" that combines the coordinates obtained by various MDs. If all MDs are well calibrated and their relative positions and orientations are measured with sufficient precision it should not create any problem.

# 3 Calibration

The calibration methods can be subdivided roughly into two classes.

Hardware calibration. With this approach the physical parameters of devices are adjusted in such a way, that variation of measurements done by calibrated device and the "golden" devise $MD_0$ be minimal.

$$\|MD_0(s) - MD_1(s)\| \rightarrow \min \tag{1}$$

Here $\|\cdot\|$ is some suitable norm.

Software calibration. In this case no physical parameters are adjusted. The calibrated device is used as it is. However, the measurements that are obtained by MD are recalculated in such a way, that they become as close as possible to the measurements obtained from the "golden" device. The same criteria (1) is used.

Software calibration can be illustrated by the following diagram (Fig. 2).

In this diagram $f_0$ and $f_1$ would represent the measurements obtained by the "golden" device $MD_0$ and the calibrated device $MD_1$. Apart from mapping $\hat{h}$, the diagram is commutative i.e. $f_0 = f_1 \circ h$. Mapping h can be called the perfect calibration since using this mapping we can make from $MD_1$ the perfect duplicate of $MD_0$. In practice, perfect calibration cannot be obtained, so that a mapping $\hat{h}$ is used to approximate h.

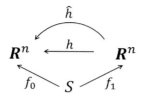

**Fig. 2** Software calibration diagram

Mapping $\hat{h}$ should satisfy the following conditions.

1. It should approximate mapping h as close as possible.
2. For any point $v \in R^n$, $\hat{h}(v)$ has to be calculated fast.

We want to note that this two conditions, generally speaking, contradict each other. The better mapping $\hat{h}$ approximate mapping $h$, the more time is necessary to calculate it. The method we offer is a compromise. We focus primarily on the speed of computation. Also we expect that our approximation will produce excellent result in the case when we use similar measurement devices since even before calibration the variation of measurements should not be too big.

We propose the following algorithm for obtaining $\hat{h}$ as $\hat{h} = g \circ u$, where u will provide barycentric coordinates and mapping g will provide interpolation with respect of certain previously chosen points. This is illustrated by the diagram below (see Fig. 3).

Note: Only lower part of the diagram is commutative.

**Algorithm**

We present the algorithm for the case $n = 3$ (MD measures location of points in 3 dimensional space). The same algorithm can be easily modified to for other values of $n$ when result of a single measurement is an array of $n$ numbers.

Chose 4 points $s_1, \ldots, s_4$. on the surface of the object S or close to it that do not lie on the same 2D plane. Let $A_1, \ldots, A_4 \in R^3$ be coordinates of these points measured by $MD_0$. Consider now $B_1, \ldots, B_4 \in R^3$ the coordinates of the same points but measured by $MD_1$.

Note: MD1 should be positioned exactly the same as $MD_0$.

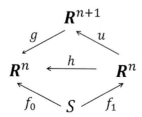

**Fig. 3** The proposed algorithm of calibration

We want $\hat{h}$ to be an affine mapping (composition of translation and linear mappings) such that $\hat{h}(B_i) = A_i$ for $i = 1, \ldots, 4$.

Let $(x, y, z)$ be the coordinates of a point $s$ as measured by $MD_1$. Denote by $B = (x, y, z)$. Now, let $\lambda_1, \ldots, \lambda_4$ be barycentric coordinates [8, 9] of $(x, y, z)$ with respect of the points $B_1, \ldots, B_4 \in R^3$. In other words $u(x, y, z) = (\lambda_1, \ldots, \lambda_4) \in R^4$.

Define $g(\lambda_1, \ldots, \lambda_4) = \lambda_1 A_1 + \cdots + \lambda_4 A_4 \in R^3$.

Barycentric coordinates of the point $B = (x, y, z)$ can be obtained by the following formulas:

$$\begin{pmatrix} \lambda_1 \\ \lambda_2 \\ \lambda_3 \end{pmatrix} = T^{-1}(B - B_4). \quad \lambda_4 = 1 - \lambda_1 - \lambda_2 - \lambda_3. \tag{2}$$

$$\text{Matrix } T = \begin{pmatrix} x_1 - x_4 & x_2 - x_4 & x_3 - x_4 \\ y_1 - y_4 & y_2 - y_4 & y_3 - y_4 \\ z_1 - z_4 & z_2 - z_4 & z_3 - z_4 \end{pmatrix} \tag{3}$$

As it can be seen from the formulas, the method has very low computational complexity. Matrices $T$ and $T^{-1}$ are calculated only once.

## 4 Conclusion

3-D modeling of various objects is very important area of research because of the numerous applications that are expected to revolutionize health care [10–12]. Additional information on medical applications for 3D printing is given in [4–6, 13–20].

To speed up the process of scanning of 3-D objects the use of several scanning devices is considered. Proposed method do not require any markers to be attached to the object. A calibration method is proposed to achieve maximum similarity among the units.

Due to the increased speed of scanning this approach has potentials for real-time applications.

**Acknowledgements** The authors would like to thank Simon Kudernatsch, Dr. Takafumi Asaki (TAMUT innovation lab) for many interesting discussions.

## References

1. Hammoudi, K., Dornaika, F., Soheilian, B., Paparoditis, N.: Extracting wire-frame models of street facades from 3D point clouds and the corresponding cadastral map. In: International

Archives of Photogrammetry, Remote Sensing and Spatial Information Sciences (IAPRS), vol. 38, part 3A, pp. 91–96, Saint-Mandé, France, 1–3 Sept 2010

2. Sitek, et al.: Tomographic reconstruction using an adaptive tetrahedral mesh defined by a point cloud. IEEE Trans. Med. Imag. **25**, 1172 (2006)
3. Bernardini, F., Rushmeier, H.E.: The 3D model acquisition pipeline (PDF). Comput. Graph. Forum. **21**(2), 149–172 (2002)
4. Ozbolat, I.T., Yu, Y.: Bioprinting toward organ fabrication: challenges and future trends. IEEE Trans. Biomed. Eng. **60**(3), 691–699 (2013). [PubMed]
5. Hoy, M.B.: 3D printing: making things at the library. Med. Ref. Serv. Q. **32**(1), 94–99 (2013)
6. D Print Exchange. National Institutes of Health; Available at: http://3dprint.nih.gov. Accessed July 9 2014
7. Boutros, P.J., Lee, C.X., Mathews, S.J., Wilkens, P.J., Peterson, D.R., McIsaac, J.H.: Novel 3-D brachial plexus reconstruction from 2-D ultrasound using XBOX Kinect tracking. Proceeding of Biomedical Engineering Society Annual Meeting, Atlanta, GA (2012)
8. Bradley, C.J.: The Algebra of Geometry: Cartesian, Areal and Projective Co-ordinates. High perception, Bath (2007). ISBN 978-1-906338-00-8
9. Coxeter, H.S.M.: Introduction to geometry, 2nd ed., pp. 216–221. Wiley, New York. ISBN 978-0-471-50458-0 (Zbl 0181.48101) (1969)
10. Lee Ventola, C.: Medical applications for 3d printing: current and projected uses. Pharm. Ther. **39**(10), 704–711 (2014)
11. Schubert, C., van Langeveld, M.C., Donoso, L.A.: Innovations in 3D printing: a 3D overview from optics to organs. Br. J. Ophthalmol. **98**(2), 159–161 (2014). [PubMed]
12. Klein, G.T., Lu, Y., Wang, M.Y.: 3D printing and neurosurgery—ready for prime time? World Neurosurg. **80**(3–4), 233–235 (2013). [PubMed]
13. Banks, J.: Adding value in additive manufacturing: researchers in the United Kingdom and Europe look to 3D printing for customization. IEEE Pulse. **4**(6), 22–26 (2013). [PubMed]
14. Mertz, L.: Dream it, design it, print it in 3-D: what can 3-D printing do for you? IEEE Pulse. **4**(6), 15–21 (2013). [PubMed]
15. Ursan, I., Chiu, L., Pierce, A.: Three-dimensional drug printing: a structured review. J. Am. Pharm. Assoc. **53**(2), 136–144 (2013). [PubMed]
16. Gross, B.C., Erkal, J.L., Lockwood, S.Y., et al.: Evaluation of 3D printing and its potential impact on biotechnology and the chemical sciences. Anal. Chem. **86**(7), 3240–3253 (2014). [PubMed]
17. Bartlett, S.: Printing organs on demand. Lancet. Respir. Med. **1**(9), 684 (2013). [PubMed]
18. Science and society: Experts warn against bans on 3D printing. Science **342**(6157), 439 (2013). [PubMed]
19. Lipson, H.: New world of 3-D printing offers "completely new ways of thinking:" Q & A with author, engineer, and 3-D printing expert Hod Lipson. IEEE Pulse. **4**(6), 12–14 (2013). [PubMed]
20. Cui, X., Boland, T., D'Lima, D.D., Lotz, M.K.: Thermal inkjet printing in tissue engineering and regenerative medicine. Recent Pat. Drug Deliv. Formul. **6**(2), 149–155 [PMC free article] [PubMed] (2012)

# More Effective Control of Linear Switched-Reluctance Motor Based on the Research of Electromagnetic Processes of Field Theory Methods Linear Electrical Machines

Pavel G. Kolpakhchyan, Alexey R. Shaikhiev
and Alexander E. Kochin

**Abstract** The results of research of electromagnetic processes in linear switched-reluctance motor of various designs are presented. The methods of the electromagnetic field theory in a three-dimensional formulation are used. The analysis of the possibility of an independent control of phases of the considered electrical machines is carried out. The technical solution to improve the accuracy of the electromagnetic force regulation is proposed.

**Keywords** Switched-reluctance motor · Electromagnetic field analysis · Electromagnetic force regulation

## 1 Introduction

One of the tendency of mobile electrical generators development on the basis of internal combustion engines is the use of a free-piston engine. The linear electric machine is used as an electromechanical transducer in such devices [1–5]. The advantage of such electric generators is the lack of switched-reluctance motor, it enhances their reliability, lets reduce weight and dimensions, adapt them to different types of fuel.

A two-stroke operating cycle is usually used in the electrical generators on the basis of a free-piston engines [3, 6, 7]. There are different types of coupling of the free-piston internal combustion engine and an electric generator [3–5]. The simplest

P.G. Kolpakhchyan (✉) · A.R. Shaikhiev
Science and Production Association "Don Technology", Novocherkassk,
Russian Federation
e-mail: kolpahchyan@mail.ru

A.E. Kochin
Russian Federation Rostov State Transport University, Rostov-on-Don,
Russian Federation

© Springer International Publishing Switzerland 2016　　　　　　　　　　43
V. Styskala et al. (eds.), *Intelligent Systems for Computer Modelling*,
Advances in Intelligent Systems and Computing 423,
DOI 10.1007/978-3-319-27644-1_5

of them is the use of a single cylinder of the internal combustion engine, when its piston is connected with the movable element of the generator. This type has not found wide application as in this case, the compression stroke must be provided by the force of electric machine. Therefore, the more common variant is the gas spring or the second cylinder of the internal combustion engine (opposite diagram) on the opposite side of the movable element. The most widespread construction is the gas spring one because it is easier in setting up and using. We will consider this variant of coupling.

## 2   Construction Features of Linear Switched-Reluctance Electrical Machine

The design elements of the electrical machine operating as an electrical generator in conjunction with a free-piston engine are in the immediate vicinity of the cylinder-piston group, in the high temperature zone. Mechanical percussive load acts on them. Therefore, the use of permanent magnets [1, 3–5] is undesirable because there is a problem of temperature conditions support and fastening of permanent magnets. One of the best options is the using of a reluctance-flux electrical machine as an electromechanical transducer working in conjunction with the free-piston engine. It has not the moving element winding and the stator winding are concentrated that simplifies the design and improves reliability. For the case under study, the reactive reluctance-flux electrical machine must be two-phase operating by turn's electrical machine.

The most common linear electrical machine is made with a cylindrical gap. The movement of the movable element is effected along the gap axis. The stator and the movable element are rotation bodies, and the stator winding has a ring shape, as illustrated in Fig. 1a (1—movable element; 2—magnetic circuit elements; 3—windings; 4—sheath). A feature of reactive reluctance-flux electrical machine is

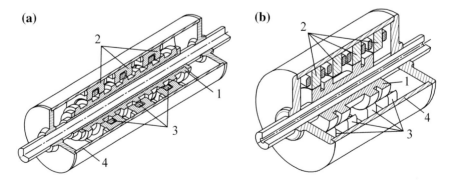

**Fig. 1** Design versions of linear reluctance-flux electrical machines

that its mass and size are determined by the size of the air gap which must be made the smallest possible [8, 9].

It imposes very high demands on the stability of the movable element position. The sliding bearings arranged at the ends of the structure perform this function in the cylindrical gap construction. Taking into account heavy mechanical loading on the movable element, the maintaining of gap stability for this design is challenging. For fixing the axial position of the movable element, the design with the longitudinal guide system is worked out, as illustrated in Fig. 1b (1—movable element; 2, 3—top an bottom part of the magnetic circuit with windings located thereon, 4—sheath).

Using magnetic circuit of three dimensional structures which consists of four poles above and below the guides is a feature of the proposed design. The outer tube (sheath) is used as the magnetic yoke which magnetic flux is closed through. Two concentrated windings are used at each phase of the electrical machine on each side of the magnetic circuit in the design under consideration.

## 3   The Influence of the Windings Connection Scheme on the Magnetic Field Distribution

The principles and methods of control of reactive reluctance-flux electrical machine have a decisive influence on its characteristics [9, 10]. Winding scheme is important for the control of the electrical machine of the proposed design. Figure 2 shows two options of winding scheme forming the magnetic fields of various configurations.

Figure 2a shows the switching-in of phase windings which creates a counter-directed magnetic fields in the lower and upper parts of the magnetic

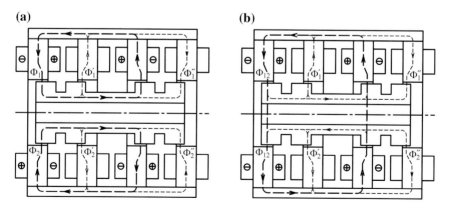

**Fig. 2** Design of the magnetic system of linear reactive reluctance-flux electrical machine and direction of the windings current of counter-directed magnetic field (**a**) and unidirectional magnetic field (**b**) in the *lower* and *upper* parts of magnetic circuit

circuit. Operational magnetic fluxes of the upper and lower winding systems $\Phi_1$ and $\Phi_2$ closed on the movable element and on the section of the sheath along the axis of the electrical machine. Besides the main magnetic flux, there are magnetic fluxes $\Phi_1'$, $\Phi_1''$ and $\Phi_2'$, $\Phi_2''$, closed through standing teeth and creating the opposite direction force. They reduce the developed power. Magnetic fluxes $\Phi_1'$ and $\Phi_2'$ through standing teeth consist of streams created by left and right pairs of windings. The magnetic flux generated by pairs of left winding passes through one standing tooth, and is distributed from the right pair of windings between the two standing teeth. In the context of the selected windings connection diagram, magnetic fluxes $\Phi_1''$ and $\Phi_2''$ equal to the remainder of fluxes through the tooth created by the left and right pairs of windings. Magnetic fluxes $\Phi_1''$ and $\Phi_2''$ are much less then fluxes $\Phi_1'$ and $\Phi_2'$, and do not lead to a significant reduction of the obtained force.

Figure 2b shows the switching-in of windings of the operating phase which create a unidirectional magnetic field in the lower and upper parts of the magnetic circuit. Unlike the previous case, in this windings connection in the working magnetic flux, $\Phi_{12}$ will be common to the upper and lower pairs of windings. The movable member and the sheath in a direction that is perpendicular to the axis of the electrical machine close it. The fluxes distribution through standing teeth has the same character as a previously described case. The difference is that in this case circuits generated through the standing teeth by left and right pairs of windings are added.

For connection scheme illustrated by Fig. 2a working magnetic flux through the upper and lower portions of the stator yoke, the movable element and sheath are closed along the axis of the electrical machine. It requires the implementation of blending of these elements parallel to this axis, which is difficult to implement. Therefore, a compound of the windings (Fig. 2b) is preferable, as the working stream is closed in the transverse direction in relation to the axis of the electrical machine. The blending of the magnetic system elements in this case is performed by simpler technology, as well as for the rotating electrical machines. Another advantage of such a winding scheme is that the magnetic fluxes pass through the standing teeth on the sliding element perpendicular to the direction of blending and virtually do not influence the electromagnetic force created by the phase.

## 4 Use of the Finite Element Method for the Electrical Machine Magnetic Field Distribution Analysis

We performed the electrical machine magnetic field distribution analysis of considered design to confirm the assumptions above. While the two-dimensional formulation of the problem it is impossible to analyze it because of the complex spatial nature of the magnetic system. So we solved the problem of the calculating the distribution of the magnetic field in the electrical machine in accordance with the principles set out in [9, 11, 12]. The analyzing of force generation of the electric

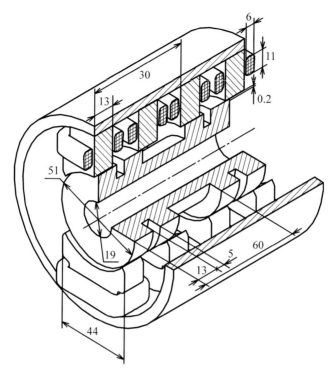

**Fig. 3** Linear reluctance-flux electrical machines domain of computation (dimensions are in millimeters)

machine depending on the position of the movable element and the characterizations of the use of active materials were the goal of our computation. We performed the computation by the finite element analysis using a software package Comsol (version 3.0).

The analysis is performed for a linear reluctance-flux electrical machines view of domain of computation and the dimensions of which are shown in Fig. 3.

In the calculations it was assumed that the stator, movable element, and sheath are made of electrical steel 50GN250. Winding wound was made of copper wire of cross-section of 2.5 mm$^2$.

Analysis of the results shows that the distribution of the magnetic field corresponds to one described above. The magnetic system saturation corresponds to the normal state of the used materials. At a small distance of the movable element from the aligned position (Fig. 4) active magnetic fluxes are much higher than standing teeth magnetic flux for both types of windings connection. It takes place because the active stator and the moving element teeth have a large overlap of the air gap, and the inactive ones have the small overlap of the air gap.

When you remove the moving element from the aligned provision (Fig. 5) active teeth overlap is reduced and inactive teeth overlap is growing. Therefore, the effect

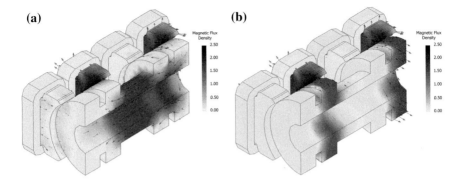

**Fig. 4** The distribution of the magnetic field in the computational domain when moving element displacement to 6 mm from the aligned position for the winding scheme making counter-directed magnetic field (**a**) and unidirectional magnetic field (**b**)

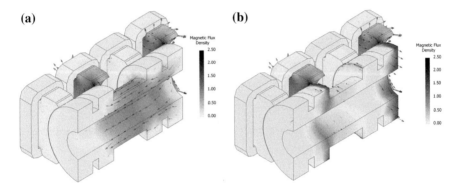

**Fig. 5** The distribution of the magnetic field in the computational domain when moving element displacement to 12 mm from the aligned position for the winding scheme making counter-directed magnetic field (**a**) and unidirectional magnetic field (**b**)

of the magnetic fluxes through inactive teeth increases. These fluxes have an inhibitory effect and reduce the electromagnetic force. This effect is shown for the windings connection scheme of the counter-directed magnetic fields (Fig. 5a), because all magnetic fluxes are closed in the direction of the magnetic circuit blending. In the case of winding connection creating a unidirectional magnetic fluxes (Fig. 5b) the worker thread closes in the direction of the magnetic blending and fluxes through inactive teeth closes wide the direction of the magnetic blending, so they have a small amount and do not influence the electromagnetic force formation.

Figure 6 shows the dependence of the electromagnetic force from the movable element moving for the schemes of the stator windings connection in question. Comparison of the results shows that of the developing electromagnetic force for both types of windings connection with little removal of the movable element from

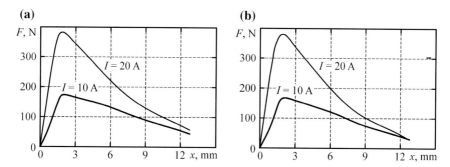

**Fig. 6** The electromagnetic force that affects the moving element depending on its displacement for the windings connections scheme creating counter-directed magnetic field (**a**) and unidirectional magnetic field (**b**)

the agreed position has no differences. The influence of the retarding magnetic flux, which is closed through inactive teeth, begins to affect with distance from the agreed position. This effect is especially affected when windings connection which creates counter-directed magnetic fluxes and almost does not affect the electromagnetic force when windings connection which creates unidirectional magnetic fluxes. It confirms conclusion mentioned above.

Currently, the authors do research on the improving the design of linear reciprocating reactive reluctance-flux electrical machine to improve its performance, reduce its weight and dimensions, find the best ways of control.

# 5 Conclusions

As follows from the analysis of current designs of electrical generators operating in conjunction with a free-piston internal combustion engine, the use of the linear reactive reluctance-flux electrical machine is sustainable. To improve the efficiency of this machine it is necessary to provide the smallest air gap. In this case, they specify strict requirements to the stability of mutual position of the stator and the movable element. It can be achieved if the longitudinal guide will be used.

The application of the stabilization system of the movable element axial position involves the use of magnetic circuit of dimensional structure. In this case, the reciprocating movement is possible when machine has two-phases operating by turns. When connecting the upper and lower windings of the operating phase the created magnetic fields passes through the standing teeth and close along the moving element. These fields lower the force on the movable element. To increase the force, it is necessary to select such current direction at left and right pairs of windings that creates flows through the standing teeth that offset each other. When the connection of the upper and lower pairs of windings, creating a magnetic field, the common working magnetic flux passes through the c upper and lower magnetic circuit. It passes perpendicularly to the axis of the electrical machine. The fluxes passing through inactive teeth are closed along the movable element. The closure of the magnetic flux along the axis of the electric machine requires a blending along this axis, it is difficult to implement in practice.

For the construction under consideration, the use of windings scheme creating a working magnetic flux perpendicular to the axis of the electrical machine is sustainable. It reduces the influence of fluxes closed through inactive teeth on the value of the electromagnetic force.

**Acknowledgements** The work is done by the authors as part of the agreement No 14.579.21.0064 about subsidizing dated 20.10.2014. The topic is "Development of experimental model of reversible electric machine reciprocating power 10–20 kW heavy-duty" by order of the Ministry of Education and Science of the Russian Federation, Federal Targeted Programme (FTP) "Researches and developments in accordance with priority areas of Russian science and technology sector evolution for 2014–2020 years". The unique identity code of applied researches (the project) is RFMEFI57914X0064.

# References

1. van Blarigan, P.: Advanced internal combustion electrical generator. In: Proceeding 2002 U.S. DOE Hydrogen Program Review, NREL/CP-610-32405, pp.1–16 (2002)
2. Mikalsen, R., Roskilly, A.P.: The fuel efficiency and exhaust gas emissions of a low heat rejection free-piston diesel engine. Proc. IMechE Part A: J. Power Energy **223**, 379–384 (2009)
3. Mikalsen, R., Roskilly, A.P.: A review of free-piston engine history and applications. Appl. Therm. Eng. **27**(14–15), 2339–2352 (2007)
4. Kosaka, H., Akita, T., Moriya, K., Goto, S. et al.: Development of free piston engine linear generator system part 1—investigation of fundamental characteristics. SAE Technical Paper 2014-01-1203 doi: 10.4271/2014-01-1203 (2014)
5. Goto, S., Moriya, K., Kosaka, H., Akita, T. et al.: Development of free piston engine linear generator system part 2—investigation of control system for generator. SAE Technical Paper 2014-01-1193 doi: 10.4271/2014-01-1193 (2014)
6. Mikalsen, R., Roskilly, A.P.: A computational study of free-piston diesel engine combustion. Appl. Energy **86**(7–8), 1136–1143 (2009)
7. Mikalsen, R., Roskilly, A.P.: The control of a free-piston engine generator. Part 1: fundamental analyses. Appl. Energy **87**, 1273–1280 (2010)
8. Vukosavic, S.N.: Electrical Machines, XXXIII, p. 649. Springer, New York (2013)
9. De Doncker, R., Pulle, D.W., Veltman, A.: Advanced Electrical Drives: Analysis, Modeling, Control. Springer, Netherlands (2011)
10. Hughes, A.: Drury, Bill. Electric Motors and Drives—Fundamentals: Types and Applications (4th Edn). Elsevier, Amsterdam (2013)
11. Mukerji, S.K., Khan, A.S., Singh, Y.P.: Electromagnetics for Electrical Machines. CRC Press, Boca Raton (2015)
12. Do, V.L., Ta, M.C.: Modeling, simulation and control of reluctance motor drives for high speed operation. In: Proceedings of "Energy Conversion Congress and Exposition, 2009. ECCE 2009. IEEE", pp. 1–6. San Jose, USA. (2009)

# Numerical Simulation of the Structural Elements of a Mobile Micro-Hydroelectric Power Plant of Derivative Type

**Denis V. Kasharin, Tatiana P. Kasharina and Michail A. Godin**

**Abstract** This article presents issues about the numerical simulation of a new mobile structure for a micro hydroelectric plant of derivative type, intended for the reserve water and electrical supply of facilities requiring low energy. The first part of the article gives a description of the elements of the derivation pipelines and their mounting conditions; The second part of the article deals with issues relating to the simulation of the stress-strain state of a single and multiple-layered flexible derivative pipeline and the simulation of its hydraulic performance using the programs Ansys Mechanical APDL and Ansys Workbench.

**Keywords** Pipeline · Composite materials · Numerical simulation · Strains solidWorks, ANSYS workbench · Static structural · Fluid CFX

## 1 Justification for the Conditions of Use of the Elements of the Mobile Hydroelectric Power Plants of Derivative Type

The construction of permanent hydroelectric power plants of derivative type is irrational, in most cases, for the decentralized electrical supply of periodical consumers with devices, which require a low energy input, using the resources of small mountain rivers. This is due to the significant expenses incurred with the setting up of derivative pipelines and structures along their tracks. The expenses are related to the construction of security structures and derivative pipelines (metal and concrete),

D.V. Kasharin (✉) · T.P. Kasharina · M.A. Godin
«Impulse» Ltd., Novocherkassk, Russia
e-mail: dendvk1@mail.ru

T.P. Kasharina
e-mail: kasharina_tp@mail.ru

M.A. Godin
e-mail: godin_m_a@mail.ru

© Springer International Publishing Switzerland 2016
V. Styskala et al. (eds.), *Intelligent Systems for Computer Modelling*,
Advances in Intelligent Systems and Computing 423,
DOI 10.1007/978-3-319-27644-1_6

51

for which heavy construction equipment and construction infrastructure are required and which causes significant impact on the environment.

Hosed micro hydroelectric plants have a lower impact on the environment. However, they have a maximal power of 16 kW and hence, the hose is not so reliable for different seasons. It gets twisted on the cross-section, and since it is placed along the banks of the small mountain watercourse, the probability of it getting damaged during flooding periods is significant.

That is why, there is a need to find a new technical solution in the form of mobile micro hydroelectric plants of derivative type, which are void of these disadvantages [1].

Since it is assumed that these micro hydroelectric plants will be used mostly non-uniformly on a daily basis for the water and power supply of facilities requiring low energy consumption, the optimal option will be pumped storage. In accordance with this scheme, part of the water from the river is pumped into the top pool, then flows through the derivation pipelines and turbines and is pumped out to the lower pool, which can function like a reservoir for clean water. When consuming low power, the micro hydroelectric plant works as a pump, pumping water from the lower reservoir to the top one, thus accumulating energy to cover peak loads.

We have currently developed a new technical solution for mobile micro hydroelectric plants of derivative type with an in-built derivation conduit made of composite material [1, 2].

The main element of this equipment is the in-built conduit, composed of a soft momentless shell made of composite materials. The top part is made of soft momentless layers for pressures of up to 50–60 m, including water or earth-filled bases.

For pressures of more than 100 m, tubes made of hard shells are used and they are calculated using known parameters.

The main difference between soft shells and hard ones is the correlation between their shapes and the loads on them. In this case, when calibrating the closed shells for water pressure, the changes in shapes due to pressure inside the shells, the fixing conditions and the hydrodynamics of the flow should be considered. The latter may affect transverse circulation which in turn can significantly affect the twisting effect in the pipelines.

In order to reduce the twisting effect under the influence of transverse circulation, provide necessary stability and work efficiency, these shells can be multi-layered.

The authors proposed the following options for derivative pipelines:

- Single-layered without base, used for slopes of less than 10° along the line for a friction coefficient of not less than 0.6;
- Single-layered with a partition without base, for slopes ranging from 10° to 15° along the line and a friction coefficient of not less than 0.6;
- Double-layered without base, where only the top layer provides adhesion;
- Double-layered with an external protective layer, which provides external protection of the pipeline.

In order to set up pipelines in areas with landslide hazards and rough terrain, it is necessary to secure the derivative pipelines properly. One way to do so is place the shell on a water-filled base, which will provide great stability and reliable efficiency.

It is required to numerically simulate the flexible pipeline structure of the mobile micro hydroelectric plant of derivative type and the mounting conditions to adjust the technical solutions. Existing analytic solutions can only give an approximate solution for the modelling of soft shells for large displacements. At the same time, the set of programs Ansys, allows to calculate models using an iterative method with geometric nonlinearity and hyper-elasticity with large strains and large deflections, taking into account hydro-dynamic effects and interactions with the contact surface [3].

Here is the sequence of numerical simulations of flexible structures of flexible pipelines for micro hydroelectric plants of derivative type.

## 2 Determining Parameters for Flexible Derivative Pipelines

### 2.1 Defining the Initial Shape of the Transverse Section of the Flexible Derivative Conduit

When defining shaping and stress-strain of the flexible derivative conduit, the following assumptions are considered: the shell is mounted on a hard, horizontal base, which is inextensible, weightless and gets filled with incompressible fluid.

Let's consider the design scheme for a water-filled soft shell (Fig. 1).

We introduce the following denotations: $W$, $B$—the width of the shell and the adhesion of the shell to the base; $p_{дн}$—the pressure at the base of the shell; $\rho g$—the specific weight of water; $L$—the perimeter of the shell; the relative adhesion of the shell to the base and the width can be determined using the following formula: $b/L$; $w = W/L$.

To determine the value of the relative adhesion of the shell, we use the formula of Meok Kim $b$ [4, 5]:

$$w \cong 0,5 p'_{дн} k^2 \left\{ 1 + k^2 \pi/32 - \left[ k^4 (3\pi - 10) \right]/64 \right\} + b \ , \tag{1}$$

where

$k$   modus of the elliptic integral;

the relative pressure on the base of the shell:

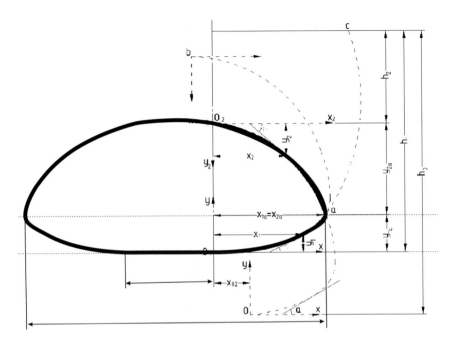

**Fig. 1** Design scheme for a water-filled soft shell

$$p'_{\text{дн}} = p_{\text{дн}}/(\rho g L).$$

To construct the upper and lower parts of the shell, we use the equation for flexible threads. For $k_1 \leq 1$ the shape of the lower thread will be an elastic of second-order, and for $k_1 > 1$—an elastic of first-order.

For the lower elastics, the parametric equation is as follows [2, 6, 7]:

$$y = h\left[1 - \sqrt{1 - k_1^2 \sin^2 \varphi}\right];$$  (2)

$$x = h\left[E(\varphi, k) - (1 - k^2/2)F(\varphi, k_1)\right],$$  (3)

where

| | |
|---|---|
| $x, y$ | are coordinates of the points of the lower thread , м; |
| $h$ | the distance of the most distant point of the lower thread to the filling surface, м; |
| $F(\varphi, k_1)$ and $E(\varphi, k_1)$ | the elastic integrals of first and second order, correspondingly; |
| $k_1$ | modus of the elliptical integral (the modus of the elastic), $k_1^2 = 4\overline{N}/h^2$; |
| $\overline{N_1}$ | the tension of the upper thread, кН/м; |

$\varphi$                      angle $\varphi = \alpha_1/2$;

$\alpha_1$              the angle between the tangent to the thread and the x-axis

Denotations $x_1$, $y_1$, $\alpha_1$, $\varphi$ at fixation points:

$$y_{1a} = h_1 \left[ 1 - \sqrt{1 - k_1^2 \sin^2 \varphi_{a1}} \right];  \tag{4}$$

$$x_{1a} = h_1 \left[ E(\varphi_{a1}, k_1) - (1 - k_1^2/2) F(\varphi_{a1}, k_1) \right],  \tag{5}$$

where $\varphi_{a1} = \alpha_{1a}/2$.

When outlining the lower thread using the Eqs. (4) and (5), the parameter $\varphi$ has a value in the range $0 \le \varphi \le \varphi_{a1}$.

For $k_1 \le 1$ the shape of the lower thread will be an elastic of second-order and for $k_1 > 1$—an elastic of first order, where the coordinates of the $x_2 O_2 y_2$ system are used for the upper thread, and the parametric equation for the upper elastic looks like the following [2, 7]:

$$y_2 = h_2 \left[ \sqrt{(1 - k_2^2 \sin^2 \varphi)/(1 - k_2^2)} - 1 \right];  \tag{6}$$

$$x_2 = h_2 \left\{ (1 - k_2^2/2)[F(\pi/2, k_2) - F(\varphi, k_2)] - [E(\pi/2, k_2) - E(\varphi, k_2)] \right\} \Big/ \sqrt{1 - k_2^2}  \tag{7}$$

where

$x_2$ and $y_2$    are the coordinates of the points of the upper thread, м;

$h_2$           the distance from the most distant point of the upper thread, м;

$k_2$           modus of the elliptic integral (modus of the elastic), $k_2^2 = 4\overline{N_2}$ $\big/ h_2^2 + 4\overline{N_2}$;

$\overline{N_2}$          tension of the upper thread, кН/м;

$\alpha_2$          the angle between the tangent to the thread and the $x_2$ axis, degrees; $\varphi_{a2} = (\pi - a_{2a})/2$.

Coordinates of the fixation point:

$$y_{2a} = h_2 \left[ \sqrt{(1 - k_2^2 \sin^2 \varphi_{a2}/1 - k_2^2)} - 1 \right];$$

$$x_{2a} = h_2 \left\{ (1 - k_2^2/2)[E_1(\pi/2, k_2) - E_1(\varphi_{a2}, k_2)] - [E_2(\pi/2, k_2) - E_2(\varphi_{a2}, k_2)] \right\} \Big/ \sqrt{1 - k_2^2}.$$

In general, when the cross-section at the fixation point is broken, $\varphi_{a2} \neq \varphi_{a1}$, we use the co-relation (1)–(7) the initial shape of the shell is determined and used for 2D simulation in the Mechanical APDL Ansys module, considering the deformation and large displacements.

## 2.2 Defining the Stress-Strain State (SSS) Considering the Properties of the Material in the Mechanical APDL

When solving tasks in Mechanical APDL, the following steps were followed:

### Selection of the type of end elements
The shells are soft and momentless and their bases are modelled correspondingly as end elements SHELL 181 and SOLID 185.

### Determining the constants of the elements, properties of the materials
The hybrid material Unisol was used as the shell material based on its breaking strength: thickness 0.004 m; density $\rho = 800$ kg/m$^3$; Poisson's coefficient to $v = 0.45$; Young modus E = $1.7 \times 10^9$ Pa. Element SOLID 185, typical for the base, has the following parameters: $\rho = 1500$ kg/m$^3$; $v = 0.45$; E = $1.7 \times 10^9$ Pa.

### Creation of a geometric model of the shells
The shape of the shells was determined according to the correlation (1)–(7), created by the models on a perimeter of 0.4 m, the internal pressure in them was from 10 to 400 kPa.

### Construction of the end-elements of the grid
These transverse outlines were used for the numerical simulation of hydraulic conditions for the performance of single—and double-chambered soft pipelines.

In order to evaluate the stability of the structures of the soft pipelines, it is necessary to determine the effects of various factors: the depth of the water in the upper pool and the friction coefficient of the bases on their deformation. Using the program ANSYS v15.0, based on the method of end-elements, the authors calculated the displacement of single and double-chambered shells for various work conditions, and as a result the parameters to obtain stability were defined.

Also, a contact couple was created between the shell and the base using CONTA 174-«contact surface», and TARGET 170-«target surface», the friction coefficient between the shell and the base were equal to 0.3, the factor of implementation tolerance FTOLN = 0.1, cohesion in the contact COHE = 0, Maximum allowed contact pressure TNOP = 1E + 20.

**Setting the boundary conditions and the load on the shell**

Since the shell is not attached, DOF Constraints were set—restrictions on the degree of freedom in all areas of the bottom surface of the base. The internal pressure is defined suing the Functions in ANSYS:

$$P = P_{\text{BHYT}} + \rho g\left(H - y_{(i)}\right),$$

where $y_{(i)}$—the coordinate of the element (i) in the y direction.

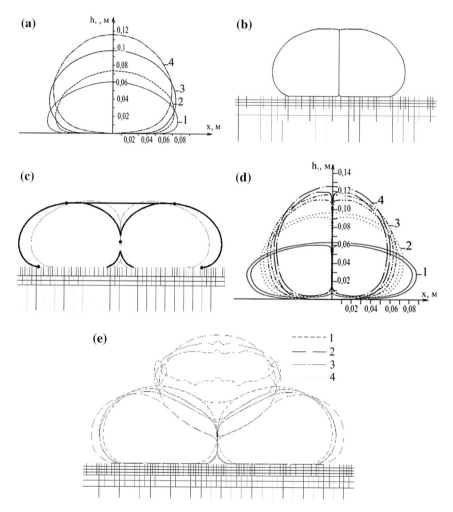

**Fig. 2** Change in the stress-strain state of a soft shell with perimeter: **a, c**—for single-chamber shell; **b**—for two-chamber shell together with the outer shell L = 0.4 m, depending on the internal overpressure in the shell (1–10 кPa, 2–50 кPa, 3–100 кPa, 4–400 кPa): **d**—water-pressured shell with a water-filled base

**Results of numerical simulation**

The end element models of shells are considered as non-linear and the calculation of their displacements is made for large displacements using the iterative method (Newton-Raphson method—the method of direct iteration). Quasi-linear equations are considered for each iteration.

The calculation results in Ansys for the transverse sections of the single-layered, single-layered with a partition, double-layered with adhesion in the top part, double-layered with external protective shell (Fig. 2) conduit, showed sufficient convergence in the case of single-layered structures with a partition and with a water-filled base, listed in the Meok Kim method, performed in FLAC [4] (Fig. 2b, c).

## 2.3 Simulating Hydrodynamic Conditions of Work for the Derivative Pipelines Using the ANSYS Workbench Module

For a comparative analysis between the hydraulic work conditions of a single-layered and a double-layered pipelines and their stress-strain state, a numerical modeling was performed in the program Ansys 15.0.

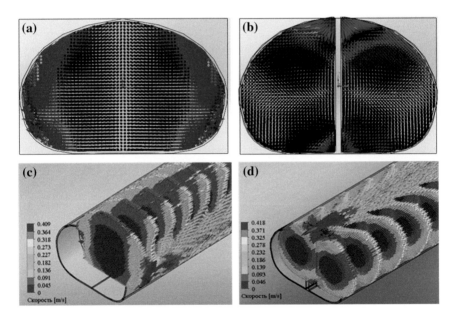

**Fig. 3** Flow dynamics in the derivative pipeline: **a**, **c**—respectively cross-section and isometry of fast structural flows in single-layered pipelines; **b**, **c**—respectively cross-section and isometry of fast structural flows in double-layered pipelines

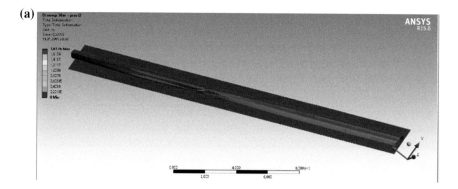

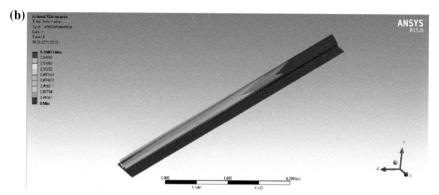

**Fig. 4** Numerical simulation of part of a pipeline of length 30 m and an inclination of 30° along its length **a, b**—single and double-layered pipelines, respectively

The models consist of part of a conduit, 30 m of length on the surface, attached to the top of a siphon outlet around the perimeter of the cross section, with a coefficient of friction between the shell and the surface of 0.3 and the angle of inclination to the horizontal plane of 30°.

A simpler case was considered with a pressure movement in the range of 1.5–3 m I the initial section of the pipe.

Single and double-layered conduits were used as initial T-sections, differing by length depending on the internal overpressure.

The initial material was an isotropic, reinforced conduit with aramid fibers, PVC. It had the following characteristics: thickness 0.004 m; density 750 kg/m$^3$; Poisson's ratio of 0.45; Young's modulus E = 1.7 × 10$^9$ Pa, a tensile strength of 1.4 MPa.

After the creation of 3-D models in SolidWorks and converting the Static Structural module in ANSYS Workbench, these forms are tested for their static state. The contact between the shell and the base was simulated using target surface, which is the base contact surface in the form of the shell. Since it is not fixed, a boundary condition is used to restrict movement of the base in all directions [8, 9].

The comparison results obtained after the analytical reference of data and the results of the numerical modeling, were satisfactory, since the deviation for horizontal and vertical projections of the cross-section of the shell for single-layered conduit was less than 8 % and for double-layered, less than 12 %.

Next, the flow of water for the same geometric model was used to calculate the velocity and the hydrodynamic pressure on the walls of the single and double-layered conduits using the Fluid CFX module. They were then re-applied as the external load on the module to determine the Static Structural deformation including those associated with the transverse twisting of the conduit (Fig. 3).

Example of results from single and double-layered pipelines with identical perimeters of external shell 0.4 m as seen in Fig. 3.

As shown in Fig. 4, at a distance of 12 m, there is twisting of the flexible single-layered conduit (Fig. 4a), which will require, if it applied, extra fixings in hardly accessible mountain conditions. At the same time a double-layered conduit, as shown in (Fig. 4b), does not show twisting and its deformations are within the limits of elastic. Therefore, despite the large amount of material used, this pipeline will be reliable for use, and the length of the sections will be increased, which will facilitate the installation of these sections.

## 3 Conclusions

1. A calculated and justified construction of pipeline was performed using composite material within the pressure range from 9.81 to 490.5 kPa. Numerical simulation using the program Ansys 15.0 in the Static Structural module showed the sufficient convergence with the results obtained analytically using formulas (1)–(7) for static calculations. Maximum deviation showed 12 %, which is acceptable for cases of large deformations. In the future, these relationships will be used for the preliminary calculation of double-layered forms of soft conduits.
2. The results of the numerical simulation of the hydraulic conditions of one- and two-chambered shells, made using the program Ansys 15.0, suggest that pipes of length over 30 m, and with an angle of inclination of 30° along the line, twist in single-layered conduits and hence, it is best to use a double-layered structure, with a water-filled base.

**Acknowledgments** The work was performed by the authors in accordance with the agreement № 14.579.21.0029 about the granting of subsidiaries dating from 05.06.2014 г. On the topic: "The development of technical solutions and technological construction of mobile micro hydroelectric power plants of derivative type for seasonal water and power supply" as a task from the Ministry of education and science of the Russian Federation FCP "Research and development in priority fields of the scientific-technological complex of Russia for the years 2014–2020". Unique identifier for the applied scientific research (Project) RFMEFI57914X0029.

# References

1. An application for an invention RF MPK E02B9/02 №2015106761. The composite cell derivation conduit and the method of its construction. Date of submission 26 Feb 2015
2. Kasharin, D.V., Kasharina, T.P., Godin, P.A., Godin, M.A.: Use of pipelines fabricated from composite materials for mobile diversion hydroelectric power plants. Power Technol. Eng. **48** (6), 448–452 (2015)
3. Nonlinear adaptive meshing and algorithms for the analysis of problems with large deformation, http://cae-expert.ru/new/nelineynoe-adaptivnoe-perestroenie-setki-i-algoritmy-dlya-analiza-zadach-s-bolshoy-deformaciey
4. Kasharin, D.V., Thai, T.K.T.: Analysis of the results of numerical simulation and experimental research of sustainable mobility dams composite, № 1, pp. 91–105 . Herald Perm National Research Polytechnic University, Construction and architecture (2014)
5. Kim, M.: Two-dimensional analysis of four types of water-filled geomembrane tubes as temporary flood-fighting devices ю M.S. Thesis, p. 202. Virginia Tech, Blacksburg (2001)
6. Kasharin, D.V.: Protective engineering structures made of composite materials in the water construction. In: Ministry of Education and Science of the Russian Federation, pp. 51–119. South-Russian State Technical University (NPI). (2012)
7. Haberyan, K.M.: Rational forms of pipes, tanks and pressure slab/Gosstroiizdat, p. 206 (1956)
8. ANSYS Mechanical APDL Structural Analysis Guide. ANSYS, Inc. pp. 841–1110 (2012)
9. ANSYS Inc. PDF Documentation for Release 15.0, http://148.204.81.206/Ansys/readme.html

# Complex System Modeling with General Differential Equations Solved by Means of Polynomial Networks

**Ladislav Zjavka and Václav Snášel**

**Abstract** Differential equations can describe physical and natural systems, which behavior only explicit exact functions are not able to model. Complex dynamic systems are characterized by a high variability of time-fluctuating data relations of a great number of state variables. Systems of differential equations can describe them but they are too unstable to be modeled unambiguously by means of standard soft computing techniques. In some cases the correct form of a differential equation might absent or it is difficult to express. Differential polynomial neural network is a new neural network type, which forms and solves an unknown general partial differential equation of an approximation of a searched function, described by discrete data observations. It generates convergent sum series of relative partial polynomial derivative terms, which can substitute for a partial or/and ordinary differential equation solution. This type of non-linear regression decomposes a system model, described by the general differential equation, into low order composite partial polynomial fractions in an additive series solution. The differential network can model the dynamics of the complex weather system, using only several input variables in some cases. Comparisons were done with the recurrent neural network, often applied for simple and solid time-series models.

**Keywords** General differential equation · Polynomial neural network · Sum relative derivative term substitution · Complex function approximation

## 1 Introduction

A lot of techniques were developed to solve differential equations and their systems. The solutions can apply sum series [4], genetic programming [6], fuzzy techniques [1], evolutionary algorithms [5] and an artificial neural network (ANN) construction

L. Zjavka (✉) · V. Snášel
VŠB—Technical University of Ostrava, IT4innovations, Ostrava
Czech Republic
e-mail: ladislav.zjavka@vsb.cz

© Springer International Publishing Switzerland 2016
V. Styskala et al. (eds.), *Intelligent Systems for Computer Modelling*,
Advances in Intelligent Systems and Computing 423,
DOI 10.1007/978-3-319-27644-1_7

[10]; they require the differential equation to be defined in a simple explicit form. ANN is able to model the non-linear nature of dynamic processes, reproduce an empirical relationship between some inputs and one or more outputs. It is applied for such purpose regarding to its approximation capability of any continuous nonlinear function with arbitrary accuracy that offer an effective alternative to more traditional statistical techniques. A common ANN operating principle is based on learned entire similarity relationships between new presented input patterns and the trained ones; however, it does not allow for eventual straight elementary data relations, which multi-variable polynomial functions can easily describe [8].

$$y = a_0 + \sum_{i=1}^{m} a_i x_i + \sum_{i=1}^{m}\sum_{j=1}^{m} a_{ij} x_i x_j + \sum_{i=1}^{m}\sum_{j=1}^{m}\sum_{k=1}^{m} a_{ijk} x_i x_j x_k + \cdots \qquad (1)$$

$m$                          number of variables $X(x_1, x_2, \ldots, x_m)$
$A(a_1, a_2, \ldots, a_m), \ldots$    vectors of parameters.

    Differential polynomial neural network (D-PNN) is a new neural network type [11], which results from the GMDH (Group Method of Data Handling) polynomial neural network (PNN), developed Ivakhnenko in 1968 [7]. It is possible to express a general connection between input and output variables by means of the Volterra functional series, a discrete analogue of which is the Kolmogorov-Gabor polynomial (1). This polynomial can approximate any stationary random sequence of observations and can be computed by either adaptive methods or a system of Gaussian normal equations. GMDH decomposes the complexity of a process, expressed by the general polynomial (1), into many simpler relationships each described by the low order polynomials (2) for every pair of the input values [9].

$$y = a_0 + a_1 x_i + a_2 x_j + a_3 x_i x_j + a_4 x_i^2 + a_5 x_j^2 \qquad (2)$$

D-PNN decomposes the general DE by means of composite functions, in a multi-layer polynomial network structure, into the 2nd order partial DE relative model (of data relations) in an additive selective term solution, analogous to the GMDH decomposition of the Kolmogorov-Gabor polynomial (1). In contrast with the common neural network functionality, each neuron (i.e. substitution derivative term) can be included directly in the total network output, which is the sum of all the selected (active) neuron output values [11]. The D-PNN application benefits should become evident in the modeling of complex dynamic systems (including e.g. weather conditions), which are possible to be described by systems of differential equations (Sect. 2). The proposed D-PNN substitution (Sects. 3 and 4) might replace the standard DE solutions for atmospheric processes; Sect. 5 gives a demonstrational example.

## 2 Numerical Meteorological Modeling

Numerical weather prediction uses mathematical models of the atmosphere and oceans to predict the weather based on current weather conditions. Fundamental problem lies in the chaotic nature of atmospheric processes, which the partial differential equations can simulate. It is impossible to solve these equations exactly, because small errors grow with time and the equations strongly interact with each other. A set of partial differential equations is solved, starting from the assumed 3D initial conditions, predicting the time-change of a variable based upon conditions in each central cell and its horizontal/vertical grid neighbors. Differential equations represent the change over an infinitesimally small spacing but the applied actual time step is only a "finite difference" of a reasonable approximation to a differential and thus larger time-steps produce larger errors. A meteorological model forms forecasts by solving the equations of motion for a fluid (3), which are derived from fundamental physical laws: conservation of momentum (Newton's laws), mass (both of air and moisture) and thermal energy (thermodynamics)[1]

$$\frac{d\vec{\mathbf{v}}}{d\mathbf{t}} = -\frac{1}{\rho}\nabla\mathbf{p} + 2\vec{\mathbf{v}} \times \vec{\Omega} + \vec{\mathbf{g}} + \vec{\mathbf{f}} \tag{3}$$

$v$  velocity vector,
$\rho$  density,
$p$  pressure,
$g$  gravidity acceleration,
$f$  vector of the friction force,
$\Omega$  angular velocity vector of Earth rotation.

The hydrostatic equilibrium equation (4) is derived from the motion equation defined for the vertical vector component (5) and static conditions $v_x = v_y = v_z = 0$. It states that the rate at which the pressure decreases with height is equal to the air density times the gravidity.

$$\frac{\partial\mathbf{p}}{\partial\mathbf{z}} = -\rho\mathbf{g} \tag{4}$$

$$\frac{\partial v_z}{\partial t} + v_x\frac{\partial v_z}{\partial x} + v_y\frac{\partial v_z}{\partial y} + v_z\frac{\partial v_z}{\partial z} = -\frac{1}{\rho}\frac{\partial p}{\partial z} - g \tag{5}$$

Continuity equation (6) expresses conservation of mass for a fluid.

---

[1]National Weather service (NWS) local observations. www.wrh.noaa.gov/mesowest/getobext. php?wfo=tfx&sid=KHLN&num=168&raw=0&dbn=m&banner=header.

$$\frac{\partial \rho}{\partial t} + \mathbf{div}(\rho \vec{v}) = 0 \tag{6}$$

Diffusion equation (7) describes diffusion of molecular processes and turbulent advection, in respect of each line coordinate.

$$\frac{\partial \mathbf{c}}{\partial t} = -\vec{v}\nabla \mathbf{c} + \frac{1}{\rho}\frac{\partial}{\partial \mathbf{x}}\left(\rho \mathbf{K_x}\frac{\partial \mathbf{c}}{\partial \mathbf{x}}\right) + \frac{1}{\rho}\frac{\partial}{\partial \mathbf{y}}\left(\rho \mathbf{K_y}\frac{\partial \mathbf{c}}{\partial \mathbf{y}}\right) + \frac{1}{\rho}\frac{\partial}{\partial \mathbf{z}}\left(\rho \mathbf{K_z}\frac{\partial \mathbf{c}}{\partial \mathbf{z}}\right) \tag{7}$$

$c$   water vapor concentration,
$K$   turbulence coefficients with respect to line coordinates.

Vorticity equation (8) is derived from motion equation (3) in respect of atmospheric processes of the air rotation around the vertical axis.

$$\frac{\mathbf{d}}{\mathbf{dt}}(\xi_r + \lambda) = -(\xi_r + \lambda)\nabla_{\mathbf{H}} \cdot \vec{v} - \vec{\mathbf{k}}\nabla_{\mathbf{H}}\alpha \times \nabla_{\mathbf{H}}\mathbf{p} + \vec{\mathbf{k}}\frac{\partial \vec{v}}{\partial \mathbf{z}} \times \nabla_{\mathbf{H}}\mathbf{v_z}$$

$$\xi_a = \xi_r + \lambda = \frac{\partial \mathbf{v_y}}{\partial \mathbf{x}} - \frac{\partial \mathbf{v_x}}{\partial \mathbf{y}} \tag{8}$$

$\xi$   vorticity,
$\alpha$   specific air volume,
$k$   base vector of the vertical axis direction.

The relative topography equation describes contours of thickness of the layer between 2 isobaric surfaces (see footnote 1).

## 3   General Differential Equation Composition

The D-PNN decomposes and substitutes for a general sum partial differential equation (9), in which an exact definition is not known in advance and which can generally describe a system model, with a sum of relative multi-variable polynomial derivative convergent term series (10).

$$a + bu + \sum_{i=1}^{n} c_i \frac{\partial u}{\partial x_i} + \sum_{i=1}^{n}\sum_{j=1}^{n} d_{ij}\frac{\partial^2 u}{\partial x_i \partial x_j} + \cdots = 0 \quad u = \sum_{k=1}^{\infty} u_k \tag{9}$$

$u = f(x_1, x_2, \ldots, x_n)$          searched function of all input variables,
$a, B(b_1, b_2, \ldots, b_n), C(c_{11}, c_{12}, \ldots)$   polynomial parameters.

Partial DE terms are formed according to the adapted method of integral analogues, which is a part of the similarity model analysis. It replaces mathematical operators and symbols of a DE by the ratio of the corresponding values. Derivatives are replaced by their integral analogues, i.e. derivative operators are removed and simultaneously along with all operators are replaced by similarly or proportion signs in equations to form dimensionless groups of variables [3].

$$u_i = \frac{\left(a_0 + a_1 x_1 + a_2 x_2 + a_3 x_1 x_2 + a_4 x_1^2 + a_5 x_2^2 + \cdots\right)^{m/n}}{b_0 + b_1 x_1 + \cdots} = \frac{\partial^m f(x_1, \ldots, x_n)}{\partial x_1 \partial x_2 \ldots \partial x_m} \quad (10)$$

$n$    combination degree of a complete polynomial of n-variables,
$m$    combination degree of denominator variables.

The fractional polynomials (10), which can substitute for the DE terms (30), describe partial relative derivative dependent changes of $n$-input variables. The numerator polynomial (10) of $n$-input variables partly defines an unknown function $u$ of Eq. (9). The denominator includes an incomplete polynomial of the competent derivative combination of variables. The root function of the numerator decreases a combination degree of the input polynomial of a term (10), in order to get dimensionless values [3]. In the case of time-series data observations an ordinary differential equation (11), which model 1-variable function, with time derivatives is solved using partial DE terms analogous to the general partial DE (9) substitution.

$$a + bf + \sum_{i=1}^{m} c_i \frac{df(t, x_i)}{dt} + \sum_{i=1}^{m} \sum_{j=1}^{m} d_{ij} \frac{df(t, x_i, x_j)}{dt} + \cdots$$
$$+ \sum_{i=1}^{m} cc_i \frac{df^2(t, x_i)}{dt^2} + \cdots = 0 \quad (11)$$

$f(t, \mathbf{x})$    function of time $t$ and independent observations $\mathbf{x}(x_1, x_2, \ldots, x_m)$.

Blocks of the D-PNN (Fig. 2) consist of neurons having the same inputs, one for each fractional polynomial derivative combination, so each neuron is considered a substitution sum DE term (10). Each block contains a single output GMDH polynomial (2), without derivative part. Neurons do not affect the block output but can participate directly in the total network output sum calculation of a DE composition. Each block has $1$ and neuron $2$ vectors of adjustable parameters $\mathbf{a}$, then $\mathbf{a}, \mathbf{b}$ (Fig. 1).

While using 2 input variables an equivalent 2nd order partial DE is solved (12), which terms involve derivatives formed with respect to all the variables applied by the GMDH polynomial (2). The D-PNN blocks form $5$ simple neurons, which can substitute for the DE terms in respect of the single $x_1, x_2$ (13) square $x_1^2, x_2^2$ (14) and combination $x_1 x_2$ (15) derivative variables of the 2nd order partial DE (12) solution, most often used to model physical or natural system non-linearities.

$$F\left(x_1, x_2, u, \frac{\partial u}{\partial x_1}, \frac{\partial u}{\partial x_2}, \frac{\partial^2 u}{\partial x_1^2}, \frac{\partial^2 u}{\partial x_1 \partial x_2}, \frac{\partial^2 u}{\partial x_2^2}\right) = 0 \quad (12)$$

where $F(x_1, x_2, u, p, q, r, s, t)$ is a function of 8 variables

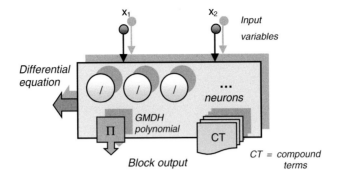

**Fig. 1** D-PNN block includes basic and compound neurons (substitution DE terms)

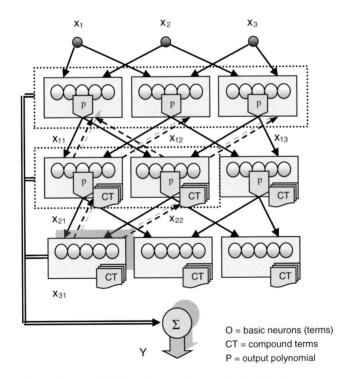

**Fig. 2** 3-variable multi-layer D-PNN with 2-variable combination blocks

$$y_1 = \frac{\partial f(x_1, x_2)}{\partial x_1} = w_1 \frac{\left(a_0 + a_1 x_1 + a_2 x_2 + a_3 x_1 x_2 + a_4 x_1^2 + a_5 x_2^2\right)^{1/2}}{1.5 \cdot (b_0 + b_1 x_1)} \tag{13}$$

$$y_4 = \frac{\partial^2 f(x_1, x_2)}{\partial x_2^2} = w_4 \frac{a_0 + a_1 x_1 + a_2 x_2 + a_3 x_1 x_2 + a_4 x_1^2 + a_5 x_2^2}{2.7 \cdot (b_0 + b_1 x_2 + b_2 x_2^2)} \tag{14}$$

$$y_5 = \frac{\partial^2 f(x_1, x_2)}{\partial x_1 \partial x_2} = w_5 \frac{a_0 + a_1 x_1 + a_2 x_2 + a_3 x_1 x_2 + a_4 x_1^2 + a_5 x_2^2}{2.3 \cdot (b_0 + b_1 x_{11} + b_2 x_{12} + b_3 x_{11} x_{12})}. \tag{15}$$

## 4   Differential Polynomial Neural Network

Multi-layer networks form composite polynomial functions (Fig. 3). Compound terms (CT), i.e. derivatives in respect of the variables of previous layer blocks, are calculated according to the composite function partial derivation rules (16) and (17). CT are formed by the products of the partial derivatives of external and internal functions.

$$F(x_1, x_2, \ldots, x_n) = f(z_1, z_2, \ldots, z_m) = f(\phi_1(X), \phi_2(X), \ldots, \phi_m(X)) \tag{16}$$

$$\frac{\partial F}{\partial x_k} = \sum_{i=1}^{m} \frac{\partial f(z_1, z_2, \ldots, z_m)}{\partial z_i} \cdot \frac{\partial \varphi_i(X)}{\partial x_k} \quad k = 1, \ldots, n \tag{17}$$

The blocks of the 2nd and following hidden layers are additionally extended with neurons, i.e. CT, which form composite partial derivatives with respect to the output and input variables of the back connected previous layer blocks, e.g. linear CT (18) and (19). The number of neurons of blocks, which involve (substitute) composite function derivatives, doubles each previous back-connected layer [11].

$$y_2 = \frac{\partial f(x_{21}, x_{22})}{\partial x_{11}} = w_2 \frac{\left(a_0 + a_1 x_{21} + a_2 x_{22} + a_3 x_{21} x_{22} + a_4 x_{21}^2 + a_5 x_{22}^2\right)^{1/2}}{1.6 \cdot x_{22}}$$
$$\cdot \frac{(x_{21})^{1/2}}{1.5 \cdot (b_0 + b_1 x_{11})} \tag{18}$$

$$y_3 = \frac{\partial f(x_{21}, x_{22})}{\partial x_1} = w_3 \frac{\left(a_0 + a_1 x_{21} + a_2 x_{22} + a_3 x_{21} x_{22} + a_4 x_{21}^2 + a_5 x_{22}^2\right)^{1/2}}{1.6 \cdot x_{22}}$$
$$\cdot \frac{(x_{21})^{1/2}}{1.6 \cdot x_{12}} \cdot \frac{(x_{11})^{1/2}}{1.5 \cdot (b_0 + b_1 x_1)} \tag{19}$$

The compound square and combination derivative terms are also calculated according to the composite function derivation rules. A combination solution may apply only some of all the potential fraction DE terms (neurons), which selection is

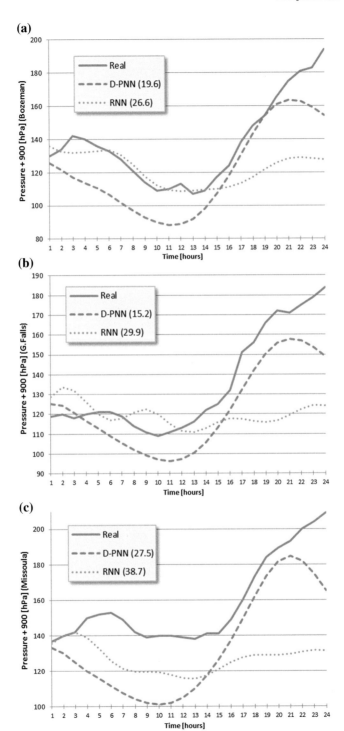

◀ **Fig. 3** 2.7.2007, conjoint static pressure time-series prediction (RMSE) in Bozeman WBAN = 24132. **a** Great Falls WBAN = 24143. **b** Missoula WBAN = 24153. **c** Montana (see footnote 2)

the principal part of a DE composition [2]. The D-PNN total output $Y$ is the arithmetical mean of all the active neuron output values (20) so as to prevent a changeable number of neurons (of a combination) from influencing the total network output sum value.

$$Y = \frac{\sum_{i=1}^{k} y_i}{k} \qquad (20)$$

$k$   actual number of active neurons

## 5 Weather Model Experiments

Numerical weather models solve the above differential equation systems (3)–(8) for a large number of state variables. Simple neural network models, using only several inputs, are able to forecast this dynamic and complex system in some cases [12]. The time-series prediction applies the previous time-step network outputs, which enter the input vector of the trained model to calculate following time estimations (21).

$$y(t) = f(y(t-1), y(t-2), \ldots, y(t-n)) \qquad (21)$$

n   number of inputs

Recurrent neural network (RNN) applies along with the input variables also its own neuron outputs from a previous time estimate, which enables it to represent a wider time-window than determined by the input vector. Both the networks RNN and D-PNN were trained with only 3 local time-series of the atmospheric pressure, the most valuable information component of meteorological data, to forecast its 24-h tendency in 3 neighboring locations of a triangle area.[2] The prediction is possible only in some chosen days as the rest of training data is highly influenced by other non-considered factors in most cases [12]. The National Oceanic and Atmospheric Administration (NOAA) provides various free historical meteorological data (see footnote 2) measured in many locations (see footnote 1).[3] 3 time hourly series of the see level pressure variables in 3 adjacent locations form 9 variables of the input vectors in total that 3 network outputs correspond (Figs. 3 and 4). The D-PNN applied only a unified structure of 3 inter-connected networks, each of them using 3 input variables (Fig. 2).

---

[2]National Climatic Data Center of National Oceanic and Atmospheric Administration (NOAA). http://cdo.ncdc.noaa.gov/qclcd_ascii/.

[3]Monthly station normals. www.ndsu.edu/ndsco/normals/7100normals/MTnorm.pdf.

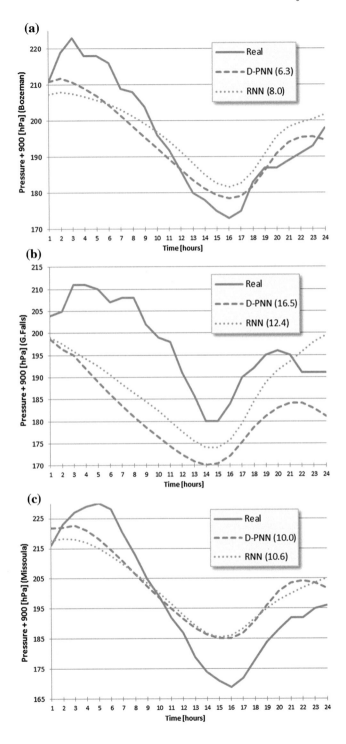

◀ **Fig. 4** 4.7.2007, conjoint static pressure time-series prediction (RMSE) in Bozeman WBAN = 24132. **a** Great Falls WBAN = 24143. **b** Missoula WBAN = 24153. **c** Montana (see footnote 2)

# 6 Conclusions

The presented new neural network type D-PNN extends the complete GMDH polynomial neural network structure to generate convergent sum series of simple and compound derivative terms, which can define and substitute for a general differential equation solution. The searched function derivative model is described by time-series data observations combined with multi-variable relations. The presented experiments show the neural network single models can forecast the very complex and uncertain weather system. The D-PNN weather model, consisting of several component networks, could form and solve a system of differential equations analogous to the described numerical meteorological model (3)–(8). The increased number of input variables might significantly impact the effect of the D-PNN applications however it requires a more complex adaptable network structure to be able to prevent from *the combinatorial explosion*, which disallows to apply all the possible 2-combination blocks. On this account the D-PNN computational complexity increases as it must select the optimal blocks (nodes) in each hidden layer along with the search of a best-fit neuron sum combination and parameter adjustment. The PNN model complexity generally increases proportionally along with the number of input variables, which is contrary to the common ANN flat 1 or 2-layer structure and this feature potentiate it to be able to solve more complicated problems of unstable systems.

**Acknowledgement** This work was supported by the IT4Innovations Centre of Excellence project (CZ.1.05/1.1.00/02.0070), funded by the European Regional Development Fund and the national budget of the Czech Republic via the Research and Development for Innovations Operational Programme and by Project SP2015/146 "Parallel processing of Big data 2" of the Student Grand System, VŠB—Technical University of Ostrava.

# References

1. Allahviranloo, T., Ahmady, E., Ahmady, N.: Nth-order fuzzy linear differential equations. Inf. Sci. **178**, 1309–1324 (2008)
2. Bertsimas, D., Tsitsiklis, J.: Simulated annealing. Stat. Sci. **8**(1), 10–15 (1993)
3. Chan, K., Chau, W.Y.: Mathematical theory of reduction of physical parameters and similarity analysis. Int. J. Theor. Phys. **18**, 835–844 (1979)
4. Chaquet, J., Carmona, E.: Solving differential equations with fourier series and evolution strategies. Appl. Soft. Comput. **12**, 3051–3062 (2012)
5. Chen, Y., Yang, B., Meng, Q., Zhao, Y., Abraham, A.: Time-series forecasting using a system of ordinary differential equations. Inf. Sci. **181**, 106–114 (2011)

6. Iba, H.: Inference of differential equation models by genetic programming. Inf. Sci. **178**(4), 4453–4468 (2008)
7. Ivakhnenko, A.: Polynomial theory of complex systems. IEEE Trans. Syst. **1**, 4 (1971)
8. Nikolaev, N.Y., Iba, H.: Polynomial harmonic GMDH learning networks for time series modeling. Neural Networks **16**, 1527–1540 (2003)
9. Nikolaev, N.Y., Iba, H.: *Adaptive Learning of Polynomial Networks*. Genetic and evolutionary computation series. Springer, New York (2006)
10. Tsoulos, I., Gavrilis, D., Glavas, E.: Solving differential equations with constructed neural networks. Neurocomputing **72**, 2385–2391 (2009)
11. Zjavka, L.: Recognition of generalized patterns by a differential polynomial neural network. Eng. Technol. Appl. Sci. Res. **2**(1), 167–172 (2012)
12. Zjavka, L.: Forecast models of partial differential equations using polynomial networks. In: *Advances in Intelligent Systems and Computing*, Vol. 238, pp. 1–11. Springer, Berlin (2013)

# Monte Carlo Probabilistic Approach Applied for Solving Problems in Mining Engineering

**K. Frydrýšek**

**Abstract** The conference paper presents some computational probabilistic solutions (i.e. direct Monte Carlo simulations and SBRA—Simulation-Based Reliability Assessment Method) applied in the branch of mining engineering. At first, the strength analysis of a bucket wheel excavator used in opencast mines for the extraction of overlying soils is solved. The assessment focuses on the arm of this particular excavator. Quasi-dynamic analysis using Finite Element Method (FEM) and measurement records of the dynamic load during the operation of a large machine can be used to derive statistical inputs (cyclic loading histograms) for probabilistic reliability assessment. At second, the solution of a hard rock (ore) disintegration process is solved (i.e. the bit moves into the ore and subsequently disintegrates it). The probabilistic results are compared with experiments and new design of excavation tool is proposed.

**Keywords** Simulation-based reliability assessment (SBRA) method · Probabilistic reliability assessment · Bucket wheel excavator · Measurements · Numerical methods · Disintegration process

## 1 Introduction

Current methods for risk/reliability assessment are mostly deterministic methods (i.e. based on constants). But in the real world, the values as loads, dimensions, material properties etc. are not fixed (constant) but variable. Finally, what's more, the factors affecting real reliability assessment are not precisely known but uncertain. In structural reliability assessment, the concept of a limit state separating a multidimensional domain of random (stochastic, probabilistic) variables into "safe" and "unsafe" domains has been generally accepted, for example see Fig. 1. For

K. Frydrýšek (✉)
Faculty of Mechanical Engineering, Department of Applied Mechanics, VSB—Technical University of Ostrava, 17. Listopadu 15/2172, 708 33 Ostrava, Czech Republic
e-mail: Karel.frydrysek@vsb.cz

© Springer International Publishing Switzerland 2016
V. Styskala et al. (eds.), *Intelligent Systems for Computer Modelling*,
Advances in Intelligent Systems and Computing 423,
DOI 10.1007/978-3-319-27644-1_8

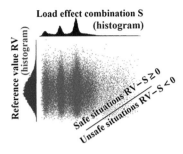

**Fig. 1** Reliability function RF (SBRA Method, result of $10^6$ Monte Carlo random simulations)

more general information about the theory, software, programming and applications of probabilistic methods see Refs. [1–8]. Stochastic methods are not widely spread in the branch of mining industry. Therefore, the applications in the mining engineering are new scientific trends.

Let us consider the Simulation-Based Reliability Assessment (SBRA) Method, a direct probabilistic Monte Carlo approach, in which all inputs are given by bounded histograms. Bounded histograms include the real variability of the inputs. Using SBRA Method, the probability of failure or an undesirable situation is obtained mainly by analysing the reliability function

$$RF = RV - S, \qquad (1)$$

See Fig. 1, where RV is the reference (allowable) value and S is a variable representing the load effect combination. The probability of failure (i.e. unsafe situation) is the probability that S exceeds RV ( i.e. $P(RF \leq 0)$). The probability of failure is a relative value depending on the definition of RV.

Hence, this paper focuses on the probabilistic numerical solution of the problems in mining engineering (i.e. the stress analysis of bucket wheel excavator used in opencast mines and the analysis of a hard rock disintegration process). Application of the SBRA Method connected with the Finite Element Method (FEM) and experiments in this area is a new and innovative trend in engineering/mining mechanics.

## 2 Probabilistic Approach to Solving Service Life of Wheel Excavator Arm (First Example—Opencast Mine)

The bucket wheel excavator, see Fig. 2, is an important large machine designed for surface extraction of large volumes of soil and minerals (i.e. overburden, lignite etc.). It usually operates on a continuous basis. The service life of such a large machine is estimated to be about 40 years. For general details on basic issues, see [4, 9–11].

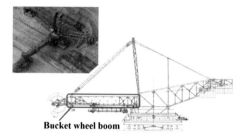

**Fig. 2** Bucket wheel excavator and its bucket wheel boom (arm)

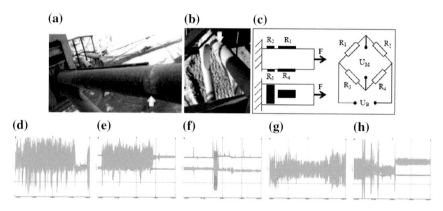

**Fig. 3** Location of the measurement points from the driver's perspective: **a** left pulling rod, **b** right pulling rod, **c** strain gauge circuit, **d–h** some examples of hourly records of measurements during operation

It is essential to measure the load of the machine. Hence, the researchers installed measuring and recording equipment on the pulling rods which are used to capture the response of both transmissions of the large excavator wheel drive. Measurements were carried out in the period of 30 days. The measuring system is composed of two strain gauge measuring circuits which monitor the relative deformations at selected locations on both rods, see Fig. 3a, b. Each rod is fitted with a strain gauge circuit composed of four glued electrical strain gauges (model 6/120 LY, made by HBM), which are connected in a full Wheatstone bridge and are weighed to the nominal zero value corresponding to the horizontal position of the bucket wheel boom, see Fig. 3c. The calibrated measurement system is also used to protect the functional parts of the machine against overload, i.e. if the limit force of $4.15 \times 10^5$ N is reached during operation. A record of the measurement was taken with a sampling frequency of 100 Hz, see [10] and Fig. 3d–h.

It is important to determine the distribution of stresses and deflections of the entire bucket wheel boom of the bucket wheel excavator during operation. Based on the results achieved, it is possible to evaluate critical points on the arm of the

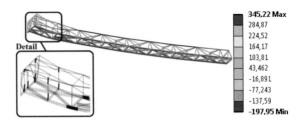

**Fig. 4** Equivalent von Mises Stresses σ (MPa) and their detail (bucket wheel boom, result of the Finite Element Method)

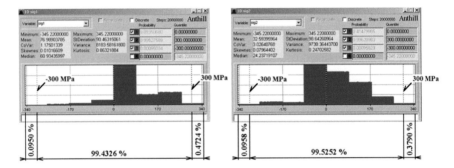

**Fig. 5** Histograms of the equivalent von Mises stresses for the bucket wheel boom in Anthill software (group A — $\sigma_A = 60.93^{+284.29}_{-406.15}$ MPa and group B — $\sigma_B = 20.24^{+320.98}_{-369.46}$ MPa)

structure and assess its safety. The solution involves discretizing the geometry of the provided drawing by means of Finite Elements (i.e. creation of a beam model) and subsequently performing a quasi-dynamic structural analysis. The calculation was performed using Ansys Workbench software, see Fig. 4.

However, the maximum equivalent stress (stress composed primarily of spatial bending and tension/compression) is then 345.22 MPa, see Fig. 4. It is apparent that the tension in certain places (near the excavator wheel) exceeds the yield strength (i.e. plasticity occurs). Thus, it is possible to reliably explain occasional fatigue fractures in the bucket wheel excavator. The stochastic analyses (i.e. statistical processing) of the first and second selected measurement records (group/file A and B), see e.g. Figs. 3d–h and 5, can be used to compile the stress histograms $\sigma_A = 60.93^{+284.29}_{-406.15}$ MPa and $\sigma_B = 20.24^{+320.98}_{-369.46}$. MPa, where the value 60.93 MPa and 20.24 MPa are the medians.

To calculate the fatigue life, it is important to determine the analysis of the size and frequency of cycles, see the differences (subtractions) between the histograms in Fig. 5 (determined by Anthill software). After computing $2 \times 10^7$ of random Monte Carlo simulations, it is possible to determine sufficiently and accurately the stress range (amplitude) $\Delta\sigma_{A,B}$ and also its absolute value $|\Delta\sigma_{A,B}|$ as

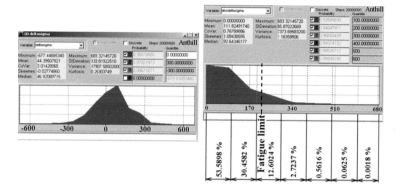

**Fig. 6** Histograms of calculated ranges of the equivalent von Mises stresses) $\Delta\sigma_{A,B} = 46.92^{+636.40}_{-724.37}$ MPa and $\left|\Delta\sigma_{A,B}\right| = 96.64^{+560.68}_{-92.64}$ MPa (Anthill software)

$$\Delta\sigma_{A,B} = \sigma_A - \sigma_B = 46.92^{+636.40}_{-724.37} \text{ MPa}, \quad \left|\Delta\sigma_{A,B}\right| = 96.64^{+560.68}_{-92.64} \text{ MPa}, \quad (2)$$

see Fig. 6. Values of 46.92 or 92.64 MPa respectively are the median values. The obtained results can be used to determine the partial or total lifetime of a bucket wheel excavator or its parts according to low-cycle and high-cycle fatigue. Due to reasons of confidentiality, the determination of the residual or total service life (based on fatigue of materials) is not presented here. However, from the real probabilistic assessment, see Fig. 6, the stress range $\left|\Delta\sigma_{A,B}\right|$ in some situations is quite high (i.e. more than fatigue limit and even more than 600 MPa). Hence, our probabilistic assessment of the remaining life of the bucket wheel excavator for the extraction of overlying soils (i.e. first example) was performed. For more information see [10].

## 3 Probabilistic Approach to Solving Hard Rock Disintegration Process (Second Example—Underground Mine)

The provision of sufficient quantities of raw materials for our civilization is one of the main limiting factors of development. Therefore, it is very important to understand the ore disintegration process, including an analysis of the bit (i.e. excavation tool) used in mining operations. The main focus is on modelling of the mechanical contact between the bit and the platinum ore and its evaluation (i.e. practical application in the mining technology), see Fig. 7. However, material properties of the ore have a large stochastic variability. Hence, the stochastic approach (i.e. SBRA Method in combination with FEM is applied). MSC. MARC/MENTAT software was used in modelling this problem. The bit moves into

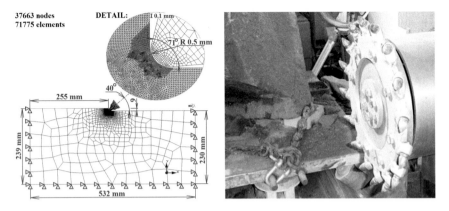

**Fig. 7** Hard rock disintegration process (FE modelling and experiment)

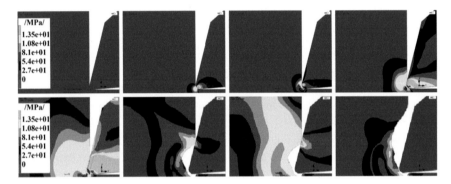

**Fig. 8** Hard rock disintegration process (equivalent von Mises stresses, FE modelling, result of 1 Monte Carlo Simulation)

the ore with the prescribed time dependent function and subsequently disintegrates it. When the bit moves into the ore (i.e. a mechanical contact occurs between the bit and the ore) the equivalent von Mises stresses in the ore increase. When the equivalent stress is greater than the tensile strength in some elements of the ore, then these elements break off. Hence, a part of the ore disintegrates. This is done by programming of deactivating the elements, see Fig. 8.

The ore material is elasto-plastic with isotropic hardening rule. The probabilistic inputs, i.e. elastic properties (Modulus of elasticity and Poisson's ratio) and plastic properties (yield stress Rp and fracture stress Rm) are described by bounded histograms, see Fig. 9a.

Because of the material nonlinearities, the mechanical contacts with friction, the large number of elements, many iteration steps, and the choice of 500 Monte Carlo simulations, four parallel computers (with 26 CPU) were used to handle the large computational requirements for this problem. The Finite Element Tearing and

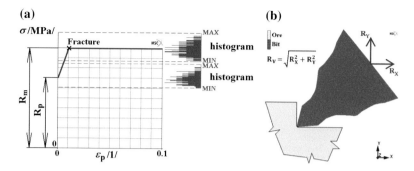

**Fig. 9 a** Stochastic material properties of the ore—stress versus plastic strain and **b** definition of the reaction force

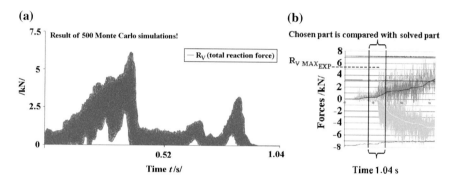

**Fig. 10** Reaction force **a** SBRA-FEM simulations and **b** experiments

Interconnect (FETI) Domain Decomposition Method (i.e. application of parallel computers, see [17]) was used.

From the results, the reaction force can be calculated. This force acts in the bit, see Figs. 9b and 10a (distribution of the total reaction forces acquired from 500 Monte Carlo simulations—stochastic result, i.e. print of 500 curves).

The calculated maximum forces (i.e. SBRA-FEM solutions) is expressed via histogram

$$Rv_{MAX_{SBRA,FEM}} = 5068^{+1098}_{-984} \, N, \qquad (3)$$

See Fig. 10a, can be compared with the experimental measurements (i.e. compared with a part of Fig. 10b). The evaluation of one force measurement, see Fig. 10b shows that the maximum measured force is

**Table 1** Second example—main acquired results

|  | Basic information | Reaction force (N) |
|---|---|---|
| Experiment | Only 1 experiment in the laboratory | 5280 |
| SBRA-FEM | 500 Monte Carlo stochastic (random) simulation | $5068^{+1098}_{-984}$ |

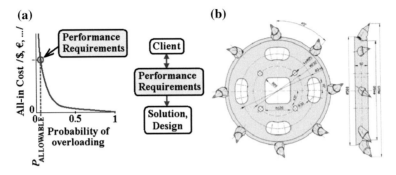

**Fig. 11 a** Definition of the acceptable probability of overloading applied for designing and **b** result of second example—final shape (i.e. design) of excavation tool for platinum ore disintegration process

$$Rv_{MAX_{EXP}} = 5280\,N. \tag{4}$$

The evaluation of SBRA-FEM simulations shows that the median value of computed force, see Eq. (3) is

$$Rv_{MAX_{SBRA,FEM_{MEDIAN}}} = 5068\,N. \tag{5}$$

Hence, the relative error calculated for the acquired median value, is

$$100 \times \frac{Rv_{MAX_{EXP}} - Rv_{MAX_{SBRA,FEM_{MEDIAN}}}}{Rv_{MAX_{EXP}}} = 4.02\,\% \tag{6}$$

The error of 4.02 % is acceptable. However, the experimental results also have large variability due to the anisotropic and stochastic properties of the material and due to the large variability of the reaction forces. The main acquired results of second example are presented in Table 1.

All the results presented in Table 1 were applied for optimizing and redesigning of the cutting bit (excavation tool), see Fig. 11. For more information see Refs. [1, 12] (i.e. more about the way of solution via parallel computers and designing).

# 4 Another Applications and Future Work

The probabilistic approaches, see [1–8, 12–16] are applied in any branch of engineering too. For example, the author of this conference paper applied them in structures on elastic foundations, see [3, 13, 14], in experiments, see [1–3, 10, 12], in biomechanics [3, 5, 12–14] etc. In structural reliability assessment, the probabilistic (i.e. stochastic) approaches have been generally accepted.

Future work is focused to finding new applications of probabilistic approaches and to improving existing numerical models.

# 5 Conclusions

1. **First example**—Based on the analysis of the load limit of a bucket wheel excavator for the extraction of overlying soils, the author determined the loads and boundary conditions for quasi-dynamic calculation using the Finite Element Method. By processing strain gauge measurement files, the researchers produced valuable histograms of stresses and their ranges using the probabilistic Simulation-Based Reliability Assessment (SBRA) Method (i.e. an application of the direct Monte Carlo stochastic approach with 2 millions of Monte Carlo random simulations solved via Anthill software). For example, the probability, when the range of $\left|\Delta\sigma_{A,B}\right| = 92.64^{+560.68}_{-92.64}$ MPa is greater than 600 MPa, is 0.0018 %. Such a high stress range has a direct destructive effect (stress exceeds the yield strength). These extreme overloads have been, and will continue to be, associated with real-life emergency situations (e.g. fatigue of materials, wheel shaft breakage, deformation of the excavator beam structure or some part of the structure; such situations are always associated with downtime and economic costs). This fact is a well-known reality in the long-term operation of large mining machines such as bucket wheel excavators. It is also hazardous when $\left|\Delta\sigma_{A,B}\right| \geq$ fatigue limit, see Fig. 6, which is certain to involve fatigue stress (i.e. when the value is probably higher than the fatigue limit or yield stress). The frequency of these cycles is about 5 %. This is therefore associated with previous and future fatigue fractures in the parts of bucket wheel excavators. The methodology presented here, using a probabilistic approach, is a new and original method for the specification and evaluation of data to determine the durability of exposed parts of bucket wheel excavators (i.e. a "real" random description of stresses and their frequencies). Further information is contained in [10]. The application of a probabilistic approach using the SBRA Method in connection with the Finite Element Method brings valuable insights into the dispersion of the real load spectrum to which the excavator is exposed. It is a highly applicable, new and original approach to solving this stochastic problem. Valuable stochastic outputs were gained from a simple beam 3D FE model in

combination with the SBRA Method as a good support for the probabilistic assessment of the remaining life of the bucket wheel excavator.

2. **Second example**—This conference paper combines the SBRA (Simulation Based Reliability Assessment) Method and FEM as a suitable tool for simulating the hard rock (ore) disintegration process too. All basic factors have been explained (i.e. 2D boundary conditions, material nonlinearities, mechanical contacts and friction between the cutting bit and the ore, the methodology for deactivating the finite elements during the ore disintegration process, application of parallel computers). The use of finite element deactivation during the ore disintegration process (as a way of programming for expanding the crack) is a modern and innovative way of solving problems of this type. The error of the SBRA-FEM results (i.e. in comparison with the experiments) is acceptable. Hence, SBRA and FEM can be a useful tool for simulating the ore disintegration process. Because the real material of the ore (i.e. yield limit, fracture limit, Young's modulus, Poisson's ratio etc.) is extremely variable, stochastic theory and probability theory were applied (i.e. programming the application of the SBRA Method). Hence, the fully probabilistic assessment was proposed according to the acceptable probability of overloading of the whole cutting-loader system. Presented results were applied for optimizing and redesigning of the cutting bit (excavation tool).

**Acknowledgements** This work has been supported by the Czech project SP2015/180.

# References

1. Frydrýšek, K., Marek, P.: Probabilistic solution and reliability assessment of the hard rock disintegration process. In: Reliability, Risk and Safety: Theory and Applications, vols. 1–3, pp. 1443-1450. CRC Press–Taylor & Francis Group, Boca Raton, USA (2010). ISBN 978-0-415-55509-8
2. Frydrýšek, K., Pečenka, L.: Probabilistic evaluation of residual stresses for hole-drilling tests. Applied Mechanics and Materials, vol 684, pp. 400–406. Trans Tech Publications, Switzerland (2014). ISSN 1660-9336. doi:10.4028/www.scientific.net/AMM.684.400
3. Frydrýšek, K., Tvrdá, K., Jančo, R., et al.: Handbook of Structures on Elastic Foundation, pp. 1–1691. VŠB—Technical University of Ostrava, Ostrava, Czech Republic (2013). ISBN 978-80-248-3238-8
4. Gottvald, J., Kala, Z.: Sensitivity analysis of tangential digging forces of the bucket wheel excavator SCHRS 1320 for different terraces. J Civil Eng Manage **18**(5), 609–620 (2012). doi:10.3846/13923730.2012.719836
5. Grepl, J., Frydrýšek, K., Penhaker, M.: A probabilistic model of the interaction between a sitting man and a seat. In: Applied Mechanics and Materials, vol 684, pp. 413–419. Trans Tech Publications, Switzerland (2014). ISSN 1660-9336. doi:10.4028/www.scientific.net/AMM.684.413
6. Marek, P., Guštar, M., Anagnos, T., et al.: Simulation-Based Reliability Assessment for Structural Engineers, p. 365. CRC Press, Boca Raton, USA (1995). ISBN 0–8493-8286-6

7. Marek, P., Brozzetti, J., Guštar, M., Tikalsky, P., et al.: Probabilistic Assessment of Structures Using Monte Carlo Simulation Background, Exercises and Software, 2nd edn. ITAM CAS, Prague, Czech Republic (2003). ISBN ISBN 80-86246-19-1
8. Kala, Z.: Sensitivity analysis of steel plane frames with initial imperfections. Eng. Struct. **33** (8), 2342–2349 (2011)
9. Bošnjak, S.M., Petković, Z.D., Simonović, A.M., Zrnić, N.D., Gnjatović, N.B.: 'Designing-in' failures and redesign of bucket wheel excavator undercarriage. Eng. Fail. Anal. **35**, 95–103 (2013). doi:10.1016/j.engfailanal.2012.12.007
10. Gondek, H., Frydrýšek, K.: Determination of Remaining Life for Bucket Wheel Excavator K 10 000—Calculation Report, pp. 1–72. VSB—Technical University of Ostrava, Ostrava, Czech Republic (2013) (written in Czech)
11. Zhi-Wei, Y., Xiao-Lei, X., Xin, M.: Failure investigation on the cracked crawler pad link. Eng. Fail. Anal. **17**, 1102–1109 (2010). doi:10.1016/j.engfailanal.2010.01.004
12. Frydrýšek, K.: Probabilistic approaches applied in the solution of problems in mining and biomechanics. In: 17th International Conference on Engineering Mechanics Engineering Mechanics 2011, pp. 151–154, ISBN 978-80-87012-33-8, Svratka, Czech Republic
13. Frydrýšek, K.: Probabilistic Calculations in Mechanics 1 (Pravděpodobnostní výpočty v mechanice 1) [CD-ROM]. VŠB—Technical University of Ostrava, Ostrava, Czech Republic (2010). ISBN 978-80-248-2314-0
14. Frydrýšek, K., Čada, R.: Probabilistic reliability assessment of femoral screws intended for treatment of "Collum Femoris" fractures. In: BIOMECHANICS 2014—International Conference of the Polish Society of Biomechanics, pp. 61–62. ISBN 978-83-7283-628-1, Łódź (2014)
15. Lokaj, A., Vavrušová, K., Rykalová, E.: Application of laboratory tests results of dowel joints in cement-splinter boards VELOX into the fully probabilistic methods (SBRA method). Applied Mechanics and Materials, vol 137, pp. 95–99 (2012). ISSN 1660-9336, doi:10.4028/www.scientific.net/AMM.137.95
16. Lokaj, A., Vavrušová, K.: Contribution to the probabilistic approach of the impact strength of wood. In: ENGINEERING MECHANICS 2011, pp. 363–366, Svratka, Czech Republic (2011). ISBN 978-80-87012-33-8
17. Farhat, C., Roux, F.X.: A method of finite element tearing and interconnecting and its parallel solution algorithm. Int. J. Numer. Meth. Eng. **32**, 1205–1227 (1991)

# Investment Funds Management Strategy Based on Polynomial Regression in Machine Learning

**Antoni Wiliński, Anton Smoliński and Wojciech Nowicki**

**Abstract** This paper presents the results of an investment strategy simulation. The strategy is based on common regression models in a time series, which yields the decision. A simple polynomial regression was the basic method used to achieve short-term value forecasts in the time series. Base params (number of steps in the past and a degree of a polynomial) were set based on a machine learning algorithm. The strategy is improved with some additional original (constitutes by the authors) parameters because using only the regression proved to be completely ineffective. Financial markets with bidirectional transactions (long and short transactions), as well as only long transaction markets, were both taken under research.

**Keywords** Time series · Prediction · Finance markets · Simulation · Algotrading · Machine learning

## 1 Introduction

To document the effectiveness of the strategy, the authors carried out some simulations on different financial markets; often on markets with entirely different characteristics and variability over time. Forex (Foreign Exchange) is one of today's most popular remote access markets to invest on. It's enabling quick and high returns on invested funds. Unfortunately it is also a high risk market, where

A. Wiliński · A. Smoliński · W. Nowicki (✉)
Faculty of Computer Science and Information Technology, West Pomeranian University of Technology Szczecin, Szczecin, Poland
e-mail: wnowicki@wi.zut.edu.pl

A. Wiliński
e-mail: awilinski@wi.zut.edu.pl

A. Smoliński
e-mail: ansmolinski@wi.zut.edu.pl

© Springer International Publishing Switzerland 2016
V. Styskala et al. (eds.), *Intelligent Systems for Computer Modelling*,
Advances in Intelligent Systems and Computing 423,
DOI 10.1007/978-3-319-27644-1_9

87

players can lose their assets very quickly. With Forex it is very difficult to provide an accurate prediction on currency pairs, this makes it difficult to trade safely. Prediction (prediction estimation before validation) should be seen as a method to react to changes in the time series. First attempts to simulate, as covered in this article, were taken on a EUR/USD currency pair as the test data to represent the financial market. This currency pair currently has the largest volume and is the most chosen financial instrument. Tests were taken on hourly OHLC bars for the period between 2200 on 17th January 2012 and 1600 on 26th June 2013, and the source of a data was bossafx.pl trading service. The OHLC chart is a representation of the financial market data as a time series where each time period is described by four prices: O—open price of a period, H—highest price in a period, L—lowest price in a period, C—close price in a period, and trading volume.

In this paper, the efficacy of the strategy for a completely different financial market such as Investment Funds was also studied. The time series for investment funds has a completely different characteristic; they are focused on long-term investment (months or even years). Authors present a completely different approach based on the same concept of polynomial regression. Data was taken from bossafund.pl online brokerage house. The data includes values for individual funds which depend on the funds from the previous day, or a few days earlier. These values (for most of funds) are updated on the portal every day from 9 am local time. Because the funds are listed on various stock exchanges, the data will not be covered continuously on all of the markets. To keep continuity of the data and avoid any issues triggered by gaps in the time series, missing data was populated using the last given value of asset.

Strategies based on the linear or the higher degree polynomial regression, as well as a multiregression, are well known methods and have a long history. Other authors, like Draper, Fujimoto, Elder and Breiman [1–4], are directly or indirectly using those methods in their research. The first author of this paper has previously published papers [5, 6] in this area. These methods used in their original form are not often effective, sometimes even completely useless. Almost always they can give good results in a short, closed time period. It's a result of a natural value found by machine learning(local optimization). Of course it could be different in a different time window of the time series. Fusion with the other prediction methods can significantly improve efficacy; In particular, good results could be achieved from a fusion of regression methods with various methods based on simple investment rules, like pattern recognition, moving average, pivot points, stop loss and take profit thresholds, and patterns from the traditional technical analysis. There is a wide range of literature concerned with simple rules in prediction on financial markets. First attempts to use these methods were taken in the 1980s, and continues to current day [7–11]. The special interest we can see in comparing risks Forex approach [12, 13] and more enhanced methods [1, 14] with low frequency series.

The authors aim is to prove that the combination of the two approaches allows us to achieve a better strategy and to find authors original solution with good results of simulation.

## 2 Characteristics of Investment Strategy

The first of the strategies for consideration (called reg1t), is based on a simple linear regression, attempting to predict time series value change in next step (i + 1). The example results of this strategy are represented by Fig. 1.

Figure 1 shows actual time series values and some linear model built on k = 11 steps back. Approximation with the linear model was taken on a segment of a time series between x = 1 and x = 11 (x is the time when sample was taken, in this case it is an hour bar on the EURUSD time series). So current time is x = 11 and the value of it is known and it is the last known value. Those values x are result of tests which allows to make conclusion that in this time series k = 11 is optimal value of a time frame in which regression model should be tested. Current real value equals 1.2813. Value of linear model equals 1.2819. Value at the next step is unknown, but basing on a model this value can be predicted. Prediction for this step is 1.2821. Based on this prediction strategy it is advancing to a long position, because the predicted value is higher than the current real value, therefore growth is assumed. After the real $y_{12}$ value is disclosed (which was unknown at the point of prediction) the decision would lead to failure, due to the real value of $y_{12}$ being unexpectedly less than the value of $y_{11}$.

Formally, a polynomial model $y_i^m$ describing values $y_i$, $i = 1, \ldots, k$ in sampling points $x_i, i = 1, \ldots, k$ can be represented by n degree polynomial:

$$y_i^m = a_n x_i^n + a_{n+1} x_i^{n-1} + \cdots + a_1 x_i + a_0, \quad \text{for } i = 1, \ldots, k. \tag{1}$$

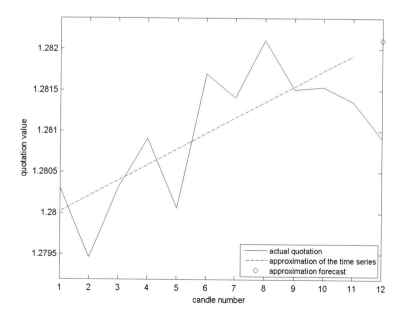

**Fig. 1** Graphical representation of the flowchart strategy *reg1t* on currency pairs market (Forex)

Vector $A = \{a_0, \ldots, a_n\}$ selection is typically carried out using the two hundred year old, least squares method to minimize the sum of a differences between $y_i$ and $y_i^m$ for $i = 1, \ldots, k$.

The parameters of this strategy are: n degree of polynomial and size of time period k. Prediction is made using a vector of factors A, calculated in a previous step and shifted by a step to determine the value of $y_{k+1}^m$ which is after the range of a sampled data.

Linear regression was determined using built-in MATLAB functions *polyfit* and *polyval*. Function *polyfit*, based on input data, is performing an interpolation and returning the factors of requested n degree polynomial. Function *polyval*, based on a given vector of a polynomial factors (form the *polyfit* function), is approximating result vector Y for a provided vector X.

Investment advice is made from a simple set of logical rules:

$$\text{if } \hat{y}_{i+1} > y_i \text{ and } a_1 > 0, \text{ then open long position} \qquad (2)$$

$$\text{if } \hat{y}_{i+1} < y_i \text{ and } a_1 < 0, \text{ then open short position} \qquad (3)$$

In those rules, factor $a_1$ is a factor of polynomial vector A placed by a linear value of the independent variable x. The second conditions, $a_1 > 0$ or $a_1 < 0$, were added to increase strategy efficiency. This new condition $a_1$ is a part of authors input into strategy based on the regression. The first condition stems from the obvious investor belief—if the model predicts an increase of value, long position is opened, as increase of price will give the profit.

The next strategy (*reg1tmetlife1*) is using the same prediction rule, the finance market is changed there to the investment fund. On an investment fund, short positions are not available. Third degree polynomial ($n = 3$) was used in regression. Degree $n = 3$ of polynomial can be a little surprise comparing to quiet efficient linear model for previous time series. Value of the param was set during attempts to find an optimum value (described as highest value of accumulated profit).

This strategy is trying to predict an investment funds value, which is a significantly slower time series than currency pair time series. The whole idea of investing in those markets is different than investing in currency or contract markets. What is particularly important to note is the fact that the value of investment fund is set after the close of a day. The results from the *reg1tmetlife1* strategy were used to build next script, *reg2tmetlife*, which is an extension on this method. Multiple funds from the Metlife family were put on the prediction. And only the best one from them is given as transaction advice.

According to the policy of investment funds, depositing funds is free of charge. In contrast to this, however, a large fee is added when investors want to withdraw their funds. This fee is often too costly that it may hinder the strategies ability to turn profit and, in some cases, could even result in some loss. Investment funds allows investors the freedom to move assets within one funds family, free of charge.

When all considered investment funds are predicting the decrease of a price, to minimize loss the investor can move all of their assets to fund with the smallest drop or withdraw them to another fund within current funds family.

## 3 Investment Strategy Tests

Each of the implemented investment strategies were taken onto parameters optimization process. The params to optimize were the steps back used to predict value, and the degree of polynomial used within regression. The optimization objective function was to maximize the gained profit for one step forward prediction. Charges like swap (on financial markets) and costs of fund management were ignored during those tests. Compared to the results, those cost don't hold any significant influence on the final result. It should be mentioned that for Forex strategies cost (difference between cost of selling and buying currency, called spread) has huge impact on a result. This has therefore been included in those strategies.

The first strategy optimization was carried out by searching the entire parameters range $k \times n$ to give a result of time period $k = 11$ and polynomial degree $n = 1$. This means that linear regression could be an efficient method to predict finance market value, limited to this market at that time.

Figure 2 shows the result of this strategy for the optimal parameters. On the abscissa is the number of tested OHLC bars. While the ordinate shows the profit

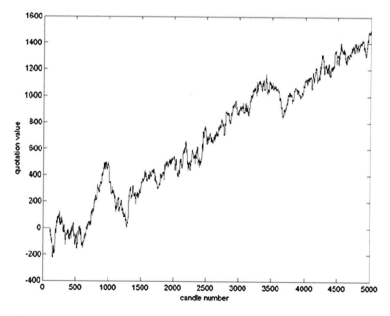

**Fig. 2** The capital accumulation curve for the strategy *reg1t* with optimal params on the test data

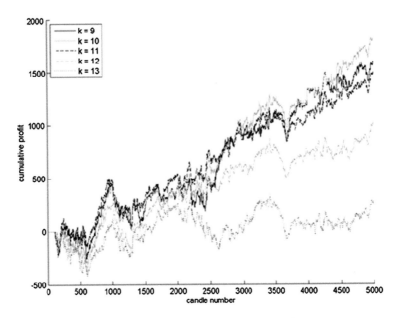

**Fig. 3** The capital accumulation curve for various values of param k

(and loss) value. Expressed in pips are the differences between purchases on an actual step, and sale price on a next step (predicted one).

To check if strategy would perform successfully for different sets of parameters, some additional tests were taken:

1. For the same time period, as in Fig. 2 and for the linear model, different values of k (number of steps back) were tested. Results are shown in Fig. 3.
2. For the static k = 11, values of n > 1 were tested. Results are shown in Fig. 4.

Results of a strategy with those parameter sets are not as promising as those of the most optimized one. However, for some curves on Figs. 3 and 4, long periods of growth could be found. This shows that some machine learning modifications should be applied to find the value of params n and k. This improved adaptive strategy was tested on a time series of the investment funds.

When regression strategy is used on investment funds market, optimal params are also expected to change. For the same time period k = 11, better results were given by third degree polynomial. Small market inertia was also an influential factor when generating results, seen in Fig. 5.

Considered strategy opened only 215 long positions using approximately 400 samples (about 18 months). This equates to transactions being taken every two days on average. This would appear to be the underlying cause for differences between this chart and the chart for Forex market. Profits are shown as a relative values (relative to fund price at the day of purchase). Final profit from, seen in Fig. 5, could be interpreted as 26 % after 500 days.

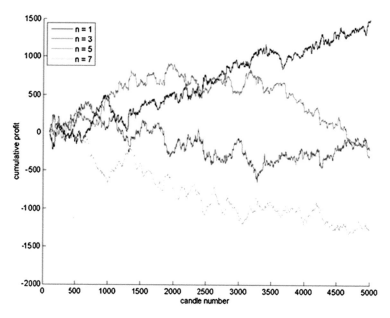

**Fig. 4** The capital accumulation curve for various values of param n

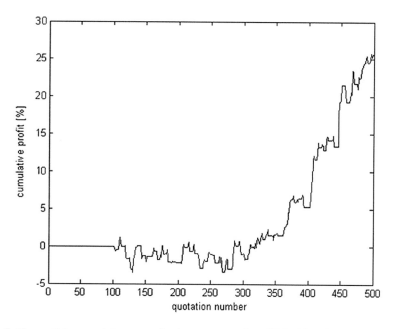

**Fig. 5** The capital accumulation curve for the strategy *reg1tmetlife1* with optimal params on the test data

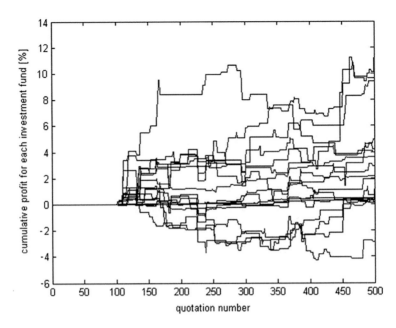

**Fig. 6** The capital accumulation curves for each of the Metlife investment fund

Next strategy implements switching between investment funds within one funds family (tested on the Metlife case). The optimal parameters for this strategy are $k = 10$ and $n = 4$. This strategy performed measurably well and produced very strong results.

Figure 6 shows the capital accumulation curves for each of the tested investment funds. It is clear that some of the funds resulted in loss, while those generating the highest profit are also characterized by significant capital drops. Results from spreading investment funds evenly to all funds (creating a fund portfolio consisting of all the funds of equal participation) are shown in Fig. 7. Experiments indicated that the diversification of funds between all investment funds within a family allows you to achieve around 3 % profit after 400 days. This is not a satisfying result. Diversification of funds allows an investor to minimize investment risk. As shown in this study, all capital loss of individual funds are compensated by profits from other funds.

The strategy depicted in Fig. 8 is completely different to previous ones; transaction is made only on investment funds which gain a profit in a last step and, if there wasn't one matching to this rule, no transaction was taken. If there was more than one with high profit funds, it was spread equally between them. During these tests, there were multiple instances where none of the investment funds had growth in the last step, there were also multiple cases where many of them had growth. After adding these improvements, results are sensational and encourage the implementation of this strategy with a substantial degree of confidence. Difficult access to funds data published in various frequencies and with different

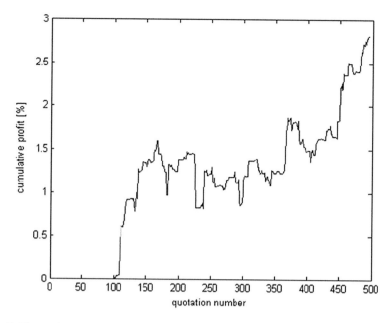

**Fig. 7** The capital accumulation curve for the portfolio of equal participation of Metlife investment funds

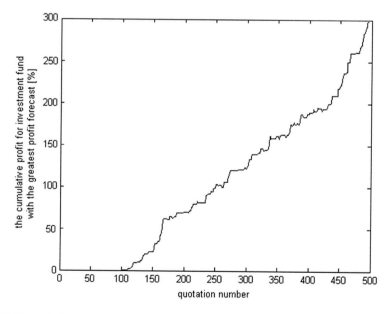

**Fig. 8** The capital accumulation curve for the portfolio of single Metlife investment fund with the highest predicted profit

delay, resulting in difficult downloading and processing, can be an obstacle to implementation.

It should also be noted that the described strategies do not include fund management costs which, depending on the individual case, can be up to 5 % of its value on an annual basis.

# 4 Conclusions

Summing up the results from this paper, it can clearly be seen that the fund market is an untapped investment instrument. Ratio between the risk and the profit is usually measured by a few different criteria, ex. Sharpe or Calmar ratio. On the Fig. 8 there is almost none of a dropdown and the highest of them is 5.46 %. From over 200 of transactions made the best single gain was 11.29 % and average profit was 0.69 % in one day. Not only does it offer long-term gain as an intermediate alternative between the bonds or the deposit and stock market investments (both in profits and risk), but it can also be used as a full-fledged speculative trading instrument. The use of automated trading strategies, although conceptual, stands in apparent contradiction with the philosophy of investment funds, but unexpectedly brings surprisingly good results.

# References

1. Elder, A.: Come Into My Trading Room. Wiley, New York (2002)
2. Draper, N.R., Smith, H.: Applied Regression Analysis. Wiley, New York (1981)
3. Fujimoto, K., Nakabayashi, S.: Applying GMDH Algorithm to Extract Rules from Examples
4. Breiman, L., Friedman, J.H., Olshen, R.A., Stone, C.J.: Classification and Regression Trees. Wadsworth International Group, Monterey (1984)
5. Wilinski, A.: Prediction models of financial markets based on multiregression algorithms. Comput. Sci. J. Mold. **19**(2), 178–188 (2011)
6. Wiliński, A.: GMDH—metoda grupowania argumentów w zadaniach zautomatyzowanej predykcji zachowań rynków finansowych. IBS PAN, Warszawa, 278 s (2009)
7. Brock, W., Lakonishok, J., LeBaron, B.: Simple technical trading rules and the stochastic properties of stock returns. J. Financ. **47**(5), 1731–1764 (1992)
8. Gencay, R.: Linear, non-linear and essential foreign exchange rate prediction with simple technical trading rules. J. Int. Econ. **47**(1), 91–107 (1999)
9. LeBaron, B.: Technical trading rules and regime shifts in foreign exchange intervention. J. Int. Econ. **49**, 125–143 (1999)
10. Tian, G., Wan, G., Guo, M.: Market efficiency and the returns to simple technical trading rules: new evidence from U.S. equity market and chinese equity markets. Asia Pac. Financ. Mark. **9**(3–4), 241–258 (2002)
11. Wiliński, A., Bera, A., Nowicki, W., Błaszyński, P.: Study on the effectiveness of the investment strategy based on a classifier with rules adapted by machine learning. ISRN Artif. Intell. **2014**, 1–10 (2014)

12. Muriel, A.: Short-term predictions in Forex trading. Phys. A Stat. Mech. Appl. **344**(1), 190–193 (2004)
13. LeBaron, B.: Technical trading rules and regime shifts in foreign exchange (1991)
14. Krutsinger, J.: Trading Systems: Secrets of the Masters. McGraw-Hill, Inc., New York (1997)

# Detecting Hidden Patterns in European Song Contest—Eurovision 2014

**Dionysios Kakouris, Georgios Theocharis, Prodromos Vlastos and Nasrullah Memon**

**Abstract** Twitter is known as a very popular tool among the users of the Internet nowadays (Larsson and Moe in J New Media and Society 14(5): 729–747, 2012 [1]), where, we share different types of opinions. In this article, we use opinion mining in order to find the winner of the Eurovision 2014—the European Song Contest. The aim of the article is to implement algorithms for opinion mining using R in order to analyze the hidden patterns in content harvested from Twitter.

**Keywords** Eurovision song contest · Hidden patterns · Microblogging · Opinion mining · Twitter

## 1 Introduction

An annual Song Contest known as the Eurovision musical competition held every year between the members of European Broadcasting Union. Blangiardo and Baio [2] mentioned in their scholarly work that the first edition of the contest was held in Lugano, Switzerland in the year 1956. Tele-voting system was introduced in the year 1997 in order to allow the viewers from the countries who are taking part in voting for their favorite act via phone, email or text.

In this article, our aim is to detect hidden patterns in the European Song Contest held in 2014 by using microblogging text from Twitter in order to extract hidden patterns from the participants. For this purpose, we use opinion mining in microblogging platform/tool. Microblogging platforms have become very popular communication tools among Internet users, huge amount of messages are posted daily in popular sites that provide services for microblogging. The authors of the messages share their opinions on a number of topics and discuss the issues of their

D. Kakouris · G. Theocharis · P. Vlastos · N. Memon (✉)
The Maersk Mc-Kinney Moller Institute, University of Southern Denmark,
Campusvej 55, 5230 Odense M, Denmark
e-mail: nasrullah.memon@gmail.com

© Springer International Publishing Switzerland 2016
V. Styskala et al. (eds.), *Intelligent Systems for Computer Modelling*,
Advances in Intelligent Systems and Computing 423,
DOI 10.1007/978-3-319-27644-1_10

importance. Therefore, microblogging platforms became a valuable source of participants' opinions and sentiments [3].

In order to create our dataset we have used microblogging website—Twitter. *Twitter is an online social network and microblogging service that enables users to send and read short 140-characters text messages, called "tweets". Twitter has seen a lot of growth since it launched in October, 2006.*[1] The users can group posts together by topic or type using hashtags. In order to retrieve tweets and create a corpus we used the hashtag #eurovision, which is an official hashtag for the Eurovision song contest. We aim to detect hidden patterns from the microblogging service, as the users' post every day their liking and disliking opinions on a number of aspects, thus, providing a treasure of valuable content. *Twitter contains an enormous number of text posts and it grows every day. The authors of these posts vary from regular users to celebrities, company representatives, politicians and so on; therefore, it is possible to collect messages of users from different social and interest groups* [3].

## 2 Literature Review

With the rise of social networks and microblogging platforms, opinion mining has been a very interesting field of study for many researchers. The authors [4] used web-blogs to make a corpus in order to mine the opinions and use emotion icons assigned to blog posts as indicators of users' mood. The authors applied Support Vector Machine (SVM) and Conditional Random Field (CRF) learners to classify opinions/sentiments for the specific sentence and then the overall sentiment of the document.

In another research [5], the author viewed the usage of emoticons such as ":-)" to form a training set for the sentiment classification. The dataset was divided into "positive" (texts with happy emoticons) and "negative" (texts with sad or angry emoticons) samples. The final result achieved an accuracy of 70 % on the test set.

A method for automatic collection of corpus was presented by Pak and Paroubek [3] that can be used to train a sentiment classifier. The authors used TreeTagger for POS-tagging and observed the difference in distributions among positive, negative and neutral sets.

Pak and Paroubek [3] used Twitter data in order to collect training data and then perform a sentiment search in a similar approach to the research carried by the author as in [6]. More specifically, the authors applied the method of sentiment analysis by examining the "positive" and "negative" emoticons and the authors achieved an accuracy of 81 % by using the Naïve Bayes classifier.

---

[1]http://www.netword.com/bargain/twitter.html.

# 3 Dataset

As mentioned in the Sect. 1, we used the hashtag #eurovision, which is an official hashtag for the Eurovision song contest. In order to be able to retrieve the tweets we created an application in the Twitter developers' website (https://dev.twitter.com/), which provides us the necessary fields (consumer key and consumer secret) and enables us to access Twitter API. In order to use R, we installed corresponding packages and libraries in order to enable to access Twitter and harvest data for further processing. After completion of the connection with the API, we retrieved the tweets and created our data set. We retrieved 1500 tweets which is the maximum number of messages that we can download with a request and we created the data frame, which consists of 740 unique users.

# 4 Opinion Mining

Hussain and Dhopte [7] mentioned that *"opinion mining has recently enjoyed a huge burst of research activity; there has been a steady undercurrent of interest for quite a while. One could count early projects on beliefs as forerunners of the area [8, 9]"*. It is possible to determine the attitude of a blogger with respect to a particular topic or to detect the polarity of given statement by a particular person or a specific news item.

In nutshell, opinion mining can mainly be used to classify the polarity of a given statement in order to discover whether the articulated opinion is positive, negative or neutral.

The author of the book *Sentiment Analysis and Opinion Mining: A comprehensive introduction and survey* [10] mentioned that *"Sentiment analysis, also called opinion mining, is the field of study that analyzes people's opinions, sentiments, evaluations, appraisals, attitudes, and emotions towards entities such as products, services, organizations, individuals, issues, events, topics, and their attributes. It represents a large problem space. There are also many names and slightly different tasks, e.g., sentiment analysis, opinion mining, opinion extraction, sentiment mining, subjectivity analysis, affect analysis, emotion analysis, review mining, etc. However, they are now all under the umbrella of sentiment analysis or opinion mining. While in industry, the term sentiment analysis is more commonly used, but in academia both sentiment analysis and opinion mining are frequently employed. They basically represent the same field of study"*. The author pointed out that the term sentiment analysis coined by the authors of the article Sentiment analysis: Capturing favorability using natural language processing [11], and the term opinion mining first appeared in the article authored by Dave et al. [12]. On the other hand, the author [10] clearly mentioned that the research on sentiments and opinions can be found in the research articles appeared earlier [13–18].

In this article, in order to effectively reveal whether a message had a positive or negative opinion about the Eurovision song contest we proceed to opinion mining/sentiment analysis dataset. The number of occurrences of positive and negative words in the corpus was counted in order to determine the documents sentiment score and as an extent of that the opinion of the users. To calculate the documents sentiment score, each positive word counts as +1 and each negative word as −1. We used a list of around 6800 words of positive opinion (2006 words) and negative opinion (4783 words) created by Hu and Liu [19] to calculate the scores for the corpus (the list is available online at http://www.cs.uic.edu/ ~ liub/ FBS/sentiment-analysis.html).

To proceed to the opinion mining task, we need to load the "plyr" and "stringr" packages, in order to define a function for sentiment score and to create a simple array with the "laply" function in order to keep the score of each word. The score function includes three parameters, the text, the words with positive sentiment and the words with the negative sentiment. In the next, we tried to clean up the text by removing punctuation, control characters (ESC, DEL) and digits and replace them with a space character. Also, we converted all the characters of the corpus to lower case and we create a list with all the words in the corpus in order to be able to compare them with the dictionaries of positive and negative words. Then we proceed to the comparison and return the score of the document and the score of each tweet.

In order to observe the sentiment score in the whole corpus we visualize the results we the use of histogram. It is clear from the Fig. 1 that the majority of the tweets are in the positive side of the histogram; therefore, we can easily conclude that the general opinion of the users of Twitter for the Eurovision song contest is positive.

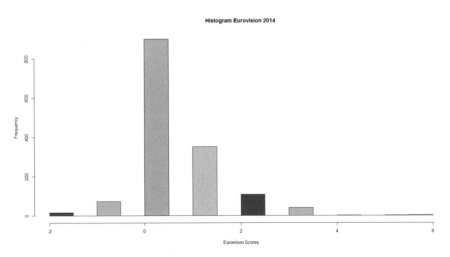

**Fig. 1** Histogram of the Tweets

# 5 Emotions and Polarity Classification

Now we decided to proceed for the classification of the tweets in terms of emotion and sentimental polarity. We classify the tweets by the emotions like: joy, surprise, anger, fear, sadness, disgust. For the polarity, we tried to discover if the message is positive, negative or neutral. We use the sentimental package in R along with the tm and Rstem package. The two functions that enable us to classify emotion and polarity are the classify_emotion function and the classify_polarity function which are included to the sentiment package.

First, by loading the packages, we prepared the data for further analysis. The preparation regards to the removal from the dataset of invalid collectors and unnecessary parts like spaces, links, punctuation and numbers. After preprocessing, we classified the emotions and categorized the polarity as shown in Figs. 2 and 3.

From Figs. 2 and 3 we can assume that the most of the tweets are joyful and the huge majority of them have a positive sentiment. Both negative and neutral sentiments/opinions are of the same level and much lower than the positive sentiment. It is therefore confirmed that the positive sentiment is much higher and that is something expected for a song contest. Usually people feel happy when talking about songs and that justified the results obtained.

After preprocessing, the data results the text in the form of string of tokens, where a token is a sequence of characters that are grouped together as a semantic unit. From that point we identified the most frequently occurring tokens in the entire corpus that can help us to detect the content of the text and information of the users' writing about in their messages. Most frequent terms are tabulated in Table 1.

Several interesting facts can be discovered from the tokens that occur with a high frequency in the corpus. We can distinguish names of countries, artists, songs and others. We will take a closer look in Tables 2, 3, 4 and 5.

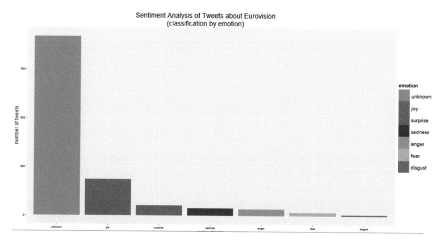

**Fig. 2** Classification of emotions

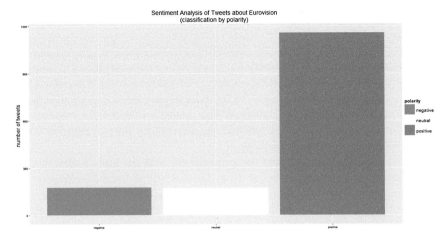

**Fig. 3** Polarity classification

**Table 1** Most frequent occurring tokens

| Frequency | Tokens |
|-----------|--------|
| 100 | esc (Eurovision Song Contest) |
| 100 | eurivi[sion] |
| 100 | joinus |
| 100 | song |
| 90 | Love |

**Table 2** Frequently occurring countries

| Frequency | Token | Countries |
|-----------|-------|-----------|
| 60 | Sweden | Sweden |
| 40 | Spain | Spain |
| 30 | Ireland | Ireland |
| 30 | Itali | Italy |
| 20 | Armenia | Armenia |
| 20 | franc[e] | France |
| 20 | Denmark | Denmark |
| 20 | Malta | Malta |

Table 2 shows that Sweden is the country with the most frequent present in the corpus, followed by Spain, and other countries. The most frequent song that exists in the messages of Twitter users is the Danish song "Cliche love song", followed from the Swedish song which also has a strong appearance in the data set as shown in Table 3. As far as the artists that exist in the data set, the most frequent present belongs to the Spanish participator Ruth Lorenzo, followed by Sanna Nielsen who is the representative of Sweden as found in Table 4. We observed that Swedish participation has intrigued twitter users as we can see its presence in all three tables

**Table 3** Frequently occurring songs

| Frequency | Token | Song | Country | Artist |
|---|---|---|---|---|
| 80 | love | Cliché love song | Denmark | Basim |
| 60 | undo | Undo | Sweden | Sanna Nielsen |
| 30 | like | Rise like a Phoenix | Austria | Carolina Wurst |
| 20 | storm | Calm after storm | Netherlands | The Common Linnets |

**Table 4** Frequently occurring artists

| Frequency | Token | Artist | Country |
|---|---|---|---|
| 80 | ruthlorenzo | Ruth Lorenzo | Spain |
| 60 | sanna, nielsen | Sanna Nielsen | Sweden |
| 50 | molli, mollysd | Molly | UK |
| 40 | Wurst | Conchita Wurst | Austria |
| 20 | Ovi | Paula Seling and Ovi | Romania |

**Table 5** Association to the token "Sweden"

| Strength of correlation | Token |
|---|---|
| 0.60 | undo (title of the song) |
| 0.58 | sanna (first name of the artist) |
| 0.53 | nielsen (surname of the artist) |
| 0.28 | ace |
| 0.16 | won |

of the countries, artists and songs, with really high frequency in all of them. Further, we observed the correlation between the tokens and found the presence of interesting associations of the token "Sweden" as it is the token with the most important presence in the corpus.

In Table 5, we can observed the tokens that are associated with the word "Sweden". As we can see the title of the song has the stronger correlation, followed from the name of the artist and also from the term "won".

In Fig. 4, we removed the word "eurovision" because its strong presence in the data set overshadowed the rest of the terms. We can distinguish some of the tokens, like 'love', as users used to write the expression 'love eurovision', 'song' as it is a song contest, 'ruthlorenzo' as it is the name of the Spanish singer, 'undo' as it is the name of the Swedish song, 'wurst' as it is the last name of the Austrian artist, and others.

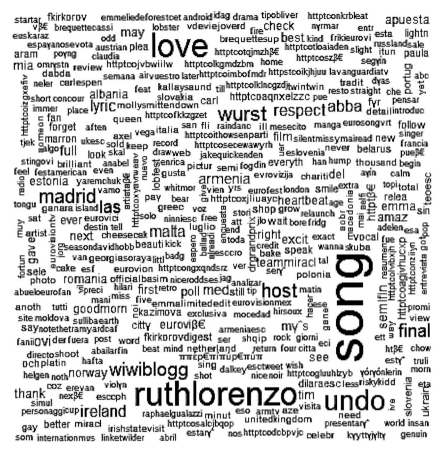

**Fig. 4** Creating a word cloud

## 6    The Second Dataset

As the days for the contest getting closer, users have a complete view of the songs and the contest, so in order to evaluate the results and be able to have a more integrated image of the user's opinion about the contest we downloaded a second data set. We retrieved tweets five days before the contest and we performed the same analysis in order to observe if there are any differences in comparison to the first data set. We observed that the users sentiment were more neutral in comparison with the first dataset, but still the overall feeling is again in the positive side. This fact is also verified by the polarity analysis as the results shown that more than 1200 tweets from the 1500 are positive and only about 200 are negative.

Emotion analysis of the second data set revealed some interesting differences. In the first data set the emotion that dominates the tweets was joy. Surprise, sadness and anger follows but in a smaller scale in comparison with joy. In the second data set surprise is the major emotion followed by joy and anger. Therefore, we observed that most of the users felt *surprised* but still felt *happy* about the contest.

As far as the frequent term analysis of the second corpus, we encountered mostly the same terms with the first data set. An interesting fact is that in the second data set, terms related to the United Kingdom participation have increased significantly and from the association analysis we can conclude that Twitter users enjoyed the song.

The implementation of this project includes several steps. First we created an application in the Twitter developer's website which provides us access to the Twitter API and therefore the ability of retrieving the tweets and creating our dataset. The second step is the performance of sentiment analysis and emotion and polarity classification of the tweets. The third step is the examination of the content of the user's messages by performing frequent term and association analysis. The final step is the repeating of the above few days before the contest in order to observe any differences and to evaluate the results.

# 7 Conclusions

Our experiments on this specific case of training data show that the results are highly dependent on the date that they were collected. Therefore, it is, safer to collect data a few days before the start of the contest in order to come into conclusions. On the other hand it is very interesting to compare older data to the latest ones, to observe the change on the tense of peoples' opinion and maybe find out why this happened. Moreover, there is a number of irrelevant information inside the training data that we do not want, that is the outlier data. For example the name of the Spanish singer (Ruth Lorenzo) was very frequently seen in our training data but that was mostly because the day that we collected the data set, it was also the voting day for the Spanish representative. The association analysis results that users link the token "Sweden" with the token "won".

From real results, we can see the winner is Austria, and in the second place we have the Netherlands and in the third place belong to Sweden. Obviously our prediction that Sweden would be most probably the winner was not correct but on the other hand Sweden finished on the top 3 countries and that gives us a good percentage of success.

It should be noted that Sweden was voted from 19 out of 26 countries and it proves the popularity that the Swedish song had. The points of the Sweden were very close to the second placed country.

As we all understand the analysis obtained through microblogging is not 100 % accurate due to valid reasons because we examined the thoughts and comments that people described which can be changed at any time for various reasons. On the other hand, we observed the tense that people have at a specific time which gives us general image of the outcome but we can only make estimations, rather than predictions.

# References

1. Larsson, Anders Olof, and Hallvard Moe. Studying political microblogging: Twitter users in the 2010 Swedish election campaign. New Media and Society. 14.5, 729–747 (2012)
2. Blangiardo, Marta, and GianlucaBaio. Evidence of bias in the Eurovision song contest: modelling the votes using Bayesian hierarchical models. Journal of Applied Statistics. 41.10, (2014)
3. Pak, E., Paroubek, P.: Twitter as a corpus for sentiment analysis and opinion mining. In: Proceedings of the Seventh Conference on International Language Resources and Evaluation 2010, Malta, May 17–23, (2010)
4. Yang, Changhua, Kevin Hsin-Yih Lin, and Hsin-Hsi Chen. Emotion classification using web blog corpora. IEEE/WIC/ACM International Conference on Web Intelligence. Fremont, CA., USA, November pp. 2–5, (2007)
5. Veeraselvi, S. J., and C. Saranya. Semantic orientation approach for sentiment classification. International Conference on Green Computing Communication and Electrical Engineering (ICGCCEE), 2014. Coimbatore, India, March pp. 6–8, (2014)
6. Read J. Using emoticons to reduce dependency in machine learning techniques for sentiment classification, in proceeding ACL student '05 Proceedings of the ACL Student Research Workshop, Association for Computational Linguistics Stroudsburg, PA, USA, pp. 43–48
7. Husain, Syed Saad, and S. V. Dhopte. An Overview of Behavioral Analaysis Using Social Web Site Data. Fifth International Conference on Communication Systems and Network Technologies (CSNT), 2015. Gwalior, MP, India, April pp. 4–6, (2015)
8. Croft, Bruce, and John Lafferty, eds. Language modeling for information retrieval. Vol. 13. Springer Science and Business Media (2013)
9. Shanahan, James G., Yan Qu, and Janyce Wiebe, eds. Computing attitude and affect in text: theory and applications. Vol. 20. Dordrecht, The Netherlands: Springer (2006)
10. Liu, Bing. Sentiment analysis and opinion mining. Synthesis Lectures on Human Language Technologies. 5.1, 1–167 (2012)
11. Nasukawa, Tetsuya, and Jeonghee Yi. Sentiment analysis: Capturing favorability using natural language processing. Proceedings of the Second international conference on Knowledge capture. Sanibel Island, FL, USA, October pp. 23–25, (2003)
12. Dave, Kushal, Steve Lawrence, and David M. Pennock. Mining the peanut gallery: Opinion extraction and semantic classification of product reviews. In Proceedings of International Conference on World Wide Web (WWW-2003), Budapest, Hungry, May pp. 20–24 (2003)
13. Das, Sanjiv, and Mike Chen. Yahoo! for Amazon: Extracting market sentiment from stock message boards. Proceedings of the Asia Pacific finance association annual conference (APFA). Vol. 35. (2001)
14. Morinaga, Satoshi, et al. Mining product reputations on the web. Proceedings of the eighth ACM SIGKDD international conference on Knowledge discovery and data mining. Edmonton, AB, Canada, July pp. 23–25, (2002)
15. Pang, Bo, Lillian Lee, and ShivakumarVaithyanathan. Thumbs up?: sentiment classification using machine learning techniques. In Proceedings of Conference on Empirical Methods in Natural Language Processing (EMNLP-2002). Philadelphia, PA, USA, July pp. 6–7, (2002)
16. Tong, Richard M. An operational system for detecting and tracking opinions in on-line discussion. In Proceedings of SIGIR Workshop on Operational Text Classification, New Orleans, Louisiana, USA, September 13, (2001)
17. Turney, Peter D. Thumbs up or thumbs down?: semantic orientation applied to unsupervised classification of reviews. In Proceedings of Annual Meeting of the Association for Computational Linguistics (ACL-2002). Stroudsburg, PA, USA

18. Wiebe, Janyce. Learning subjective adjectives from corpora. In Proceedings of National Conf. on Artificial Intelligence (AAAI-2000), Austin, Texas, USA, July 30-August 3, (2000)
19. Hu, Minqing and Bing Liu. Mining and summarizing customer reviews. In Proceedings of ACM SIGKDD International Conference on Knowledge Discovery and Data Mining (KDD-2004), Seattle, WA, USA, August 22 –25, (2004)

# Numerical Solution of Volterra Linear Integral Equation of the Third Kind

**Taalaybek Karakeev, Dinara Rustamova**
**and Zhumgalbubu Bugubayeva**

**Abstract** In this work we consider Volterra linear integral equation of the third kind in the case when the external known function reduces to zero in the internal point of integration interval. The solution of it exists in the space of continuous functions. The numerical solution is constructed and its convergence to the solution of the initial equation is proved.

**Keywords** Volterra integral equation · Numerical solution · Fixed point iteration

## 1  Introduction

The problems of uniqueness and stability of the solution of various types of Volterra integral equations of the third kind are investigated in [1–7]. Among the works devoted to Volterra integral equations of the third kind it is possible to note work [2] in which existence of multiple parameter solutions of Volterra linear integral equations of the third kind is proved. The regularization problems of Volterra linear integral equations of the third kind are studied in [1, 8] with non-increasing external function, which reduces to zero only at the end of integration interval. In [4, 5] this function is non-decreasing and reduces to zero at the beginning of an integration interval. Regularization of this problem for the external function, which degenerates at two points, is considered in [6].

Justification of a numerical solution of the problem with non-increasing external function, on the basis of the regularized equation and on the right rectangles

T. Karakeev (✉) · D. Rustamova · Z. Bugubayeva
Zh. Balasagyn Kyrgyz National University, Bishkek, Kyrgyzstan
e-mail: tkarakeev@yandex.ru

D. Rustamova
e-mail: drustamova@list.ru

Z. Bugubayeva
e-mail: tbugubaeva@mail.ru

© Springer International Publishing Switzerland 2016
V. Styskala et al. (eds.), *Intelligent Systems for Computer Modelling*,
Advances in Intelligent Systems and Computing 423,
DOI 10.1007/978-3-319-27644-1_11

111

quadrature formulas is given in [1, 9–11]. This method is applied to the problem for external non-decreasing function in [5].

In this work as method of the final sums the approximate solution of Volterra linear integral equations of the third kind in a case when known function out of integral reduces to zero in the internal point of integration interval.

Thus there will be a possibility of use of property of Volterra integral equation and on corresponding parts of a segment of integration to apply lemmas and theorems proved for the equations with non-increasing and non-decreasing functions out of integral.

Let's assume that there exists the solution of Volterra linear integral equation of the third kind

$$p(x)\varphi(x) + \int_0^x K(x,t)\varphi(t)dt = g(x), \tag{1}$$

in $C[0,b]$.

Let the known functions $p(x), K(x,t), g(x)$ are subject to the following conditions:

$$p(x) = \begin{cases} p_1(x), x \in [0,b_1], \\ p_2(x), x \in [b_1,b], \end{cases} \quad g(x) = \begin{cases} g_1(x), x \in [0,b_1], \\ g_2(x), x \in [b_1,b], \end{cases}$$

(a) $p_1(x) \in C^2[0,b_1], g_1(x) \in C[0,b_1], p_1(x)$ − non-increasing function,

(b) $p_2(x), g_2(x) \in C^2[b_1,b], p_2(x)$ − non-decreasing function,

  $p_1(x)|_{x=b_1} = 0, \quad p_2(x)|_{x=b_1} = 0, \quad g_1(b_1) = g_2(b_1)$

(c) $K(x,t) \in C(D), \quad K(x,x) \geq 0, \quad D = \{(x,t)/0 \leq t \leq x \leq b\};$

(d) $G(x) \geq d_1, \quad G(x) = C_0 p(x) + K(x,x), \quad 0 < C_0, d_1 = const.$

If $x \in [0,b_1]$, then we obtain from Eq. (1)

$$p_1(x)u(x) + \int_0^x K(x,t)u(t)dt = g_1(x), \quad x \in [0,b_1]. \tag{2}$$

At $x \in [b_1,b]$ Eq. (1) reduces to equation

$$p_2(x)v(x) + \int_{b_1}^x K(x,t)v(t)dt = g_2(x) - \int_0^{b_1} K(x,t)u(t)dt, \quad x \in [b_1,b]. \tag{3}$$

Considering condition $g_1(b_1) = g_2(b_1)$ at $x = b_1$ from (2) we will obtain $g_2(b_1) - \int_0^{b_1} K(b_1,t)u(t)dt = 0$. From these assumptions it follows the fulfillment of matching condition $u(b_1) = v(b_1)$.

We integrate Eq. (2) and summate the obtained expression with the initial Eq. (2). Then we will obtain the equation which is equivalent to (2)

$$p_1(x)u(x) + \int\limits_0^x G_1(t)u(t)dt = \int\limits_0^x L(x,t)u(t)dt + \mu_1(x), \tag{4}$$

where $L(x,t) = K(t,t) - K(x,t) - C_0 \int\limits_s^t K(s,t)ds, \quad 0 < C_0 = const,$

$$G_1(x) = C_0 p_1(x) + K(x,x), \quad \mu_1(x) = g_1(x) + C_0 \int\limits_0^x g_1(t)dt.$$

We construct regularization of (4) in the form

$$(\varepsilon + p_1(x))u_\varepsilon(x) + \int\limits_0^x G_1(t)u_\varepsilon(t)dt = \int\limits_0^x L(x,t)u_\varepsilon(t)dt + \mu_1(x) + \varepsilon u(0), \tag{5}$$

where $\varepsilon \in (0,1)$.

Let us write (5) in the form

$$
\begin{aligned}
u_\varepsilon(x) = &-\frac{1}{\varepsilon + p_1(x)} \int\limits_0^x \exp\left(-\int\limits_t^x \frac{G_1(s)}{\varepsilon + p_1(s)}ds\right) \frac{G_1(t)}{\varepsilon + p_1(t)} \left[\int\limits_0^t [K(s,s) - K(t,s)]\right. \\
&\times u_\varepsilon(s)ds - C_0 \int\limits_0^t \int\limits_s^t K(v,s)u_\varepsilon(s)dvds - \int\limits_0^x [K(s,s) - K(x,s)]u_\varepsilon(s)ds \\
&\left. -C_0 \int\limits_0^x \int\limits_s^x K(v,s)u_\varepsilon(s)dvds\right] dt + \frac{1}{\varepsilon + p_1(x)} \exp\left(-\int\limits_0^x \frac{G_1(s)}{\varepsilon + p_1(s)}ds\right) \left[\int\limits_0^x [K(s,s)\right. \\
&\left. - K(x,s)]u_\varepsilon(s)ds - C_0 \int\limits_0^x \int\limits_s^x K(v,s)u_\varepsilon(s)dvds\right] - \frac{1}{\varepsilon + p_1(x)} \\
&\times \int\limits_0^x \exp\left(-\int\limits_t^x \frac{G_1(s)}{\varepsilon + p_1(s)}ds\right) \frac{G_1(t)}{\varepsilon + p_1(t)} [\mu_1(x) - \mu_1(t)]dt + \frac{1}{\varepsilon + p_1(x)} \\
&\times \exp\left(-\int\limits_0^x \frac{G_1(s)}{\varepsilon + p_1(s)}ds\right) \mu_1(x) + \frac{\varepsilon}{\varepsilon + p_1(x)} \exp\left(-\int\limits_0^x \frac{G_1(s)}{\varepsilon + p_1(s)}ds\right) u_h(0).
\end{aligned}
\tag{6}
$$

Let $\omega_h^1 = \{x_i = ih, i = 0\ldots n, b_1 = nh\}$—a uniform grid of $[0, b_1]$, $C_h$—the space of grid functions $u_i = u(x_i)$.

We approximate integrals in (6) using a right rectangles quadrature formula. We obtain the system of linear algebraic equations

$$u_{\varepsilon,i} = -\frac{1}{\varepsilon+p_{1,i}}h\sum_{j=1}^{i}\exp\left(-h\sum_{k=j+1}^{i}\frac{G_{1,k}}{\varepsilon+p_{1,k}}\right)\frac{G_{1,j}}{\varepsilon+p_{1,j}}\left[h\sum_{k=1}^{j-1}\left(K_{k,k}-K_{j,k}\right)u_{\varepsilon,k}\right.$$

$$-C_0h\sum_{k=1}^{j-1}h\sum_{m=k+1}^{j-1}K_{m,k}u_{\varepsilon,k}-h\sum_{k=1}^{i-1}\left(K_{k,k}-K_{i,k}\right)u_{\varepsilon,k}-C_0h\sum_{k=1}^{i-1}h\sum_{m=k+1}^{i-1}K_{m,k}u_{\varepsilon,k}$$

$$+\mu_{1,j}-\mu_{1,i}\Bigg]+\frac{1}{\varepsilon+p_{1,i}}\exp\left(-h\sum_{k=1}^{i}\frac{G_{1,k}}{\varepsilon+p_{1,k}}\right)\left[h\sum_{k=1}^{i-1}\left(K_{k,k}-K_{i,k}\right)u_{\varepsilon,k}\right.$$

$$-C_0h\sum_{k=1}^{i-1}h\sum_{m=k+1}^{i-1}K_{m,k}u_{\varepsilon,k}+\mu_{1,i}+\varepsilon u_0\Bigg],\quad i=1\ldots n,$$

$$(7)$$

where $L_{i,k}=L(x_i,x_k),u_{\varepsilon,k}=u_\varepsilon(x_k),\mu_{1,i}=\mu_1(x_i),p_{1,i}=p_1(x_i),x_j=jh$.

In [1, p. 84] the following theorem is proved.

**Theorem 1** Let the conditions $(a-d)$ take place and $\varepsilon=O(h^\alpha)$. Then, for all $0<\alpha\le\frac{1}{2}$, the solution of the system (7) as $h\to0$ uniformly converges to the exact solution of Eq. (1) $\varphi_i$ and the estimate takes place

$$\left\|u_{\varepsilon,i}-u_i\right\|_{C_h}\le C_2h^\alpha,\quad 0<C_2=const.$$

As previously, we reduce (3) to equivalent equation

$$p_2(x)v(x)+\int_{b_1}^{x}G_2(t)v(t)dt=\int_{b_1}^{x}L(x,t)v(t)dt+\mu_2(x),\quad x\in[b_1,b],\qquad(8)$$

where $G_2(t)=C_0p_2(t)+K(t,t)$,

$$\mu_2(x)=g_2(x)+\int_{0}^{b_1}K(x,t)u(t)dt+C_0\int_{b_1}^{x}g_2(t)dt+C_0\int_{b_1}^{x}\int_{0}^{b_1}K(t,s)u(s)dsdt.$$

For (8) we construct the equation with small parameter $\varepsilon\in(0,1)$ by the following rule

$$(\varepsilon+p_2(x))v_\varepsilon(x)+\int_{b_1}^{x}G_2(t)v_\varepsilon(t)dt=\int_{b_1}^{x}L(x,t)v_\varepsilon(t)dt+\mu_2(x)+\varepsilon v_h(b_1).\quad(9)$$

Using the resolvent of the kernel $\left(-\frac{G_2(t)}{\varepsilon+p_2(x)}\right)$ Eq. (9) reduces to the form, similar to Eq. (6) in which instead of $u_\varepsilon(x),p_1(x),\mu_1(x),u(0)$ will stand $v_\varepsilon(x),p_2(x),\mu_2(x),v_h(b_1)$.

In interval $[b_1,b]$ we introduce a uniform grid $\omega_h^2=\{x_i=b_1+ih,i=0\ldots k,b-b_1=kh\}$, where $k\in N$.

Using the right rectangles quadrature formula for integrals in (10), we will receive system of the linear algebraic equations

$$
\begin{aligned}
\upsilon_{\varepsilon,i} = & -\frac{1}{\varepsilon + p_{2,i}} h \sum_{j=1}^{i} \exp\left(-h \sum_{k=j+1}^{i} \frac{G_{2,k}}{\varepsilon + p_{2,k}}\right) \frac{G_j}{\varepsilon + p_{2,j}} \left[ h \sum_{k=1}^{j-1} (K_{k,k} - K_{j,k}) \upsilon_{\varepsilon,k} \right. \\
& - C_0 h \sum_{k=1}^{j-1} h \sum_{m=k+1}^{j-1} K_{m,k} \upsilon_{\varepsilon,k} - h \sum_{k=1}^{i-1} (K_{k,k} - K_{i,k}) \upsilon_{\varepsilon,k} - C_0 h \sum_{k=1}^{i-1} h \sum_{m=k+1}^{i-1} K_{m,k} \upsilon_{\varepsilon,k} \\
& + \mu_{2,j} - \mu_{2,i} \bigg] + \frac{1}{\varepsilon + p_{2,i}} \exp\left(-h \sum_{k=1}^{i} \frac{G_{2,k}}{\varepsilon + p_{2,k}}\right) \left[ h \sum_{k=1}^{i-1} (K_{k,k} - K_{i,k}) \upsilon_{\varepsilon,k} \right. \\
& - C_0 h \sum_{k=1}^{i-1} h \sum_{m=k+1}^{i-1} K_{m,k} \upsilon_{\varepsilon,k} + \mu_{2,i} + \varepsilon \upsilon_{h,0} \bigg], \quad i = 1 \ldots k, \quad \upsilon_{h,0} = u_{\varepsilon,n}.
\end{aligned}
$$

$$(10)$$

Here $\upsilon_{\varepsilon,m} = \upsilon_{\varepsilon}(x_m), \mu_{2,i} = \mu_2(x_i), p_{2,i} = p_2(x_i), i = 0, 1 \ldots k$.

For further investigation we need estimates of the following lemmas which proofs are given in [1, p. 66]

**Lemma 1** If the conditions $(a - d)$ satisfied with $\varepsilon \in (0, 1), 0 < \sigma \le \frac{1}{2}$ the following inequality takes place

$$
\left| \int_0^{x_i} \frac{G(v)dv}{\varepsilon + p(v)} - h \sum_{m=1}^{i} \frac{G_m}{\varepsilon + p_m} \right| \le C_1 h^{\sigma}, \sigma = 1 - 2\alpha, \quad i = 1 \ldots k, 0 < C_1 = const.
$$

**Lemma 2** Let the conditions $(a - d)$ satisfied and $\upsilon(x) \in C^1[0, b]$. Then there exists such a number $0 < N_4$, that the estimate takes place

$$
\left\| H_{\varepsilon}^{h}[\omega_i] \right\|_{C_h} \le N_4, \quad 0 < N_4 = const,
$$

where action of the operator $H_{\varepsilon}^{h}$ on the net function $\omega_0, \omega_1, \ldots, \omega_n$ is determined by a formula

$$
\begin{aligned}
H_{\varepsilon}^{h}[\omega_i] = & -\frac{1}{\varepsilon + p_i} h \sum_{j=1}^{i} \exp\left(-h \sum_{m=j+1}^{i} \frac{G_m}{\varepsilon + p_m}\right) \frac{G_j}{\varepsilon + p_j} (\omega_j - \omega_i) \\
& + \frac{1}{\varepsilon + p_i} \exp\left(-h \sum_{m=1}^{i} \frac{G_m}{\varepsilon + p_m}\right) (\omega_i - \omega_0).
\end{aligned}
$$

**Theorem 2** If the conditions $(a - d)$ are satisfied and $\varepsilon \in (0, 1)$, for all $0 < \sigma \le \frac{1}{2}$, the solution of system (10) at $h \to 0$ uniformly converges to the exact solution of Eq. (3) $\upsilon_i$ and the following estimate takes place

$$\left\| v_{\varepsilon,i} - v_i \right\|_{C_h} \leq N_5 h^{\alpha}, \quad 0 < N_5 = const.$$

**Proof** Using the substitution $v_{\varepsilon}(x) = v(x) + \eta_{\varepsilon}(x)$, from (5) we obtain the equation

$$(\varepsilon + p_2(x))\eta_{\varepsilon}(x) + \int_{b_1}^{x} G_2(t)\eta_{\varepsilon}(t)dt = \int_{b_1}^{x} L(x,t)\eta_{\varepsilon}(t)dt - \varepsilon v(x) + \varepsilon v_h(b_1). \quad (11)$$

We reduce Eq. (11), using a kernel rezolventa $\left( -\dfrac{G_2(t)}{\varepsilon + p_2(x)} \right)$ in the following form

$$
\eta_{\varepsilon}(x) = -\frac{1}{\varepsilon + p_2(x)} \int_0^x \exp\left( -\int_t^x \frac{G_2(s)}{\varepsilon + p_2(s)}ds \right) \frac{G_2(t)}{\varepsilon + p_2(t)} \left[ \int_0^t [K(s,s) - K(t,s)] \right.
$$

$$
\times \eta_{\varepsilon}(s)ds - C_0 \int_0^t \int_s^t K(v,s)v_{\varepsilon}(s)dvds - \int_0^x [K(s,s) - K(x,s)]\eta_{\varepsilon}(s)ds
$$

$$
\left. -C_0 \int_0^x \int_s^x K(v,s)\eta_{\varepsilon}(s)dvds \right]dt + \frac{1}{\varepsilon + p_2(x)}\exp\left( -\int_0^x \frac{G_2(s)}{\varepsilon + p_2(s)}ds \right)\left[ \int_0^x [K(s,s) \right.
$$

$$
\left. - K(x,s)]\eta_{\varepsilon}(s)ds - C_0 \int_0^x \int_s^x K(v,s)\eta_{\varepsilon}(s)dvds \right] - \frac{\varepsilon}{\varepsilon + p_2(x)}
$$

$$
\times \int_0^x \exp\left( -\int_t^x \frac{G_2(s)}{\varepsilon + p_2(s)}ds \right) \times \frac{G_2(t)}{\varepsilon + p_2(t)}[v(x) - v(t)]dt + \frac{\varepsilon}{\varepsilon + p_2(x)}
$$

$$
\times \exp\left( -\int_0^x \frac{G_2(s)}{\varepsilon + p_2(s)}ds \right)v(x) + \frac{\varepsilon}{\varepsilon + p_2(x)}\exp\left( -\int_0^x \frac{G_2(s)}{\varepsilon + p_2(s)}ds \right)v_h(b_1).
$$

$$(12)$$

Taking $x = x_i, i = 1 \ldots k$ in (12), we will apply a formula of the right rectangles to integrals in this equation. Then we obtain the system

$$
\eta_{\varepsilon,i} = -\frac{1}{\varepsilon + p_{2,i}} h \sum_{j=1}^{i} \exp\left( -h \sum_{k=j+1}^{i} \frac{G_{2,k}}{\varepsilon + p_{2,k}} \right) \frac{G_{2,j}}{\varepsilon + p_{2,j}} \left[ h \sum_{k=1}^{j-1} (K_{k,k} - K_{j,k})\eta_{\varepsilon,k} \right.
$$

$$
- C_0 h \sum_{k=1}^{j-1} h \sum_{m=k+1}^{j-1} K_{m,k}\eta_{\varepsilon,k} - h \sum_{k=1}^{i-1} (K_{k,k} - K_{i,k})\eta_{\varepsilon,k} - C_0 h \sum_{k=1}^{i-1} h \sum_{m=k+1}^{i-1} K_{m,k}\eta_{\varepsilon,k}
$$

$$
\left. + \varepsilon(v_j - v_i) \right] + \frac{1}{\varepsilon + p_{2,i}}\exp\left( -h \sum_{k=1}^{i} \frac{G_{2,k}}{\varepsilon + p_{2,k}} \right)\left[ h \sum_{k=1}^{i-1} (K_{k,k} - K_{i,k})\eta_{\varepsilon,k} \right.
$$

$$
\left. - C_0 h \sum_{k=1}^{i-1} h \sum_{m=k+1}^{i-1} K_{m,k}\eta_{\varepsilon,k} + \varepsilon(v_i - v_h(b_1)) \right] + R_i, \quad i = 1 \ldots k,
$$

$$(13)$$

where $R_i$—the sum of all residual members of integrals.

Using the estimates we obtain

$$\left|\eta_{\varepsilon,i}\right| = \left| -\frac{1}{\varepsilon+p_{2,i}} h \sum_{j=1}^{i} \exp\left(-h \sum_{k=j+1}^{i} \frac{G_{2,k}}{\varepsilon+p_{2,k}}\right) \frac{G_{2,j}}{\varepsilon+p_{2,j}} \left[ h \sum_{k=1}^{j-1} \left(K_{k,k}-K_{j,k}\right)\eta_{\varepsilon,k} \right.\right.$$

$$- C_0 h \sum_{k=1}^{j-1} h \sum_{m=k+1}^{j-1} K_{m,k}\eta_{\varepsilon,k} - h \sum_{k=1}^{i-1} \left(K_{k,k}-K_{i,k}\right)\eta_{\varepsilon,k} - C_0 h \sum_{k=1}^{i-1} h \sum_{m=k+1}^{i-1} K_{m,k}\eta_{\varepsilon,k}$$

$$+ \varepsilon\left(v_j - v_i\right) \right] + \frac{1}{\varepsilon+p_{2,i}} \exp\left(-h \sum_{k=1}^{i} \frac{G_{2,k}}{\varepsilon+p_{2,k}}\right) \left[ h \sum_{k=1}^{i-1} \left(K_{k,k}-K_{i,k}\right)\eta_{\varepsilon,k} \right.$$

$$\left.\left.- C_0 h \sum_{k=1}^{i-1} h \sum_{m=k+1}^{i-1} K_{m,k}\eta_{\varepsilon,k} + \varepsilon(v_i - v_h(b_1)) \right] \right| + \varepsilon\left|H_\varepsilon^h(v_i)\right| + |R_i|$$

$$\leq T_{12} \sum_{k=1}^{i-1} \left|\eta_{\varepsilon,k}^h\right| + \varepsilon\left|H_\varepsilon^h(v_i)\right| + |R_i|, \quad i = 1\ldots k,$$

where $0 < T_{12} = const.$

Applying difference analog of Gronwall-Bellman lemma, we have

$$\left|\eta_{\varepsilon,i}\right|_{C_h} \leq \left(\varepsilon\left|H_\varepsilon^h(v_i)\right| + |R_i|\right)\exp(T_{12}b_1).$$

Then, due to the estimate from lemma 2, on the grid norm we obtain

$$\left\|\eta_{\varepsilon,i}\right\|_{C_h} \leq \left(\varepsilon\left|H_\varepsilon^h(v_i)\right| + N_6\frac{h}{\varepsilon} + N_7 h\right)\exp(T_{12}b_1)$$

$$\leq \left(\varepsilon N_4 + N_6\frac{h}{\varepsilon} + N_7 h\right)\exp(T_{12}b_1), \quad 0 < N_6, \quad N_7 = const.$$

Therefore, considering the relation $\varepsilon = O(h^\alpha), 0 < \sigma \leq \frac{1}{2}$ we prove the Theorem 2.

Solution of Eq. (1) in the grid knots $\omega_h = \omega_h^1 \times \omega_h^2$ is determined by the following rule

$$\varphi(x_i) = \begin{cases} u(x_i), & x_i \in \omega_h^1, \\ v(x_i), & x_i \in \omega_h^2. \end{cases}$$

Let the grid function $\varphi_{\varepsilon,i}$ is defined by

$$\varphi_{\varepsilon,i} = \varphi_\varepsilon(x_i) = \begin{cases} u_\varepsilon(x_i), & x_i \in \omega_h^1, \\ v_\varepsilon(x_i), & x_i \in \omega_h^2, \end{cases}$$

where $u_\varepsilon(x_i), v_\varepsilon(x_i)$—are corresponding solutions of Eqs. (5), (8).

**Theorem 3** If conditions ($a - d$) are satisfied and $\varepsilon = O(h^\alpha)$ for all $0 < \alpha \le \frac{1}{2}$ then at $h \to 0$  $\varphi_{\varepsilon,i}$ uniformly converges to the exact solution of Eq. (1) $\varphi_i$

$$\left\| \varphi_{\varepsilon,i} - \varphi_i \right\|_{C_h} \le N_5 h^\alpha, \quad 0 < N_5 = const.$$

If the condition ($d$) is violated, then the condition ($e$) takes place

(e)  $B(x) \ge d_2$, $B(x) = C_0 p(x) + K(x,x) + C_1 g(x)$,   $0 < C_1$, $d_2 = const.$

In this case we suggest the following scheme of numerical solution of Eq. (1).

We act on Eq. (2) with operator $I + C_0 J + C_1 T$, where $I$—an unity operator, $J$ and $T$—Volterra operators of the forms

$$(Jv)(x) = \int_0^x v(t)dt, \quad (Tv)(x) = \int_0^x u(t)v(t)dt,$$

and we will make regularization

$$(\varepsilon + p_1(x))u_\varepsilon(x) + \int_0^x B_1(t)u_\varepsilon(t)dt = \int_0^x L(x,t)u_\varepsilon(t)dt + C_1 \int_0^x p_1(t)u_\varepsilon^2(t)dt$$

$$+ C_1 \int_0^x u_\varepsilon(t)dt \int_t^x K(s,t)u_\varepsilon(s)ds + \mu_1(x), \quad \varepsilon \in (0,1). \tag{14}$$

The approximate solution of Eq. (14) in grid knots $\omega_h^1$ is defined by a following rule

$$u_{\varepsilon,i} = -\frac{h}{\varepsilon + p_{1,i}} \sum_{j=1}^{i-1} \exp\left(-h \sum_{k=j+1}^{i} \frac{B_{1,k}}{\varepsilon + p_{1,k}}\right) \frac{B_{1,j}}{\varepsilon + p_{1,j}} \left\{ h \sum_{k=1}^{j-1} \left[ L_{j,k} - L_{i,k} \right] u_{\varepsilon,k} \right.$$

$$- h \sum_{k=j}^{i-1} L_{i,k} u_{\varepsilon,k} - C_1 h \sum_{k=j+1}^{i} p_{1,k} u_{\varepsilon,k}^2 - C_1 h \sum_{k=1}^{j-1} u_{\varepsilon,k} h \sum_{m=j}^{i} K_{m,k} u_{\varepsilon,m} - C_1 h \sum_{k=j}^{i-1} u_{\varepsilon,k}$$

$$\times h \sum_{m=k+1}^{i} K_{m,k} u_{\varepsilon,m} + \mu_j - \mu_i \right\} + \frac{1}{\varepsilon + p_{1,0}} \exp\left(-h \sum_{k=1}^{i} \frac{B_{1,k} + p_k'}{\varepsilon + p_{1,k}}\right) \left\{ h \sum_{j=1}^{i-1} L_{i,j} u_{\varepsilon,j} \right.$$

$$+ C_1 h \sum_{j=1}^{i} p_{1,j} u_{\varepsilon,j}^2 + C_1 h \sum_{j=1}^{i-1} u_{\varepsilon,j} h \sum_{k=j+1}^{i} K_{k,j} u_{\varepsilon,k} + \mu_{1,i} \right\}, \quad i = 1 \ldots n.$$

$$\tag{15}$$

For Eq. (3) also as well as above in grid knots $\omega_h^2$ system of the algebraic equations similar to system (15).

**Remark** A system of the algebraic Eqs. (15) is system of nonlinear algebraic equations to which solution it is possible to apply a method of successive approximations.

## 2   Conclusion

Follows from the stated theoretical part that the method of the final sums on the basis of methods of small parameter and quadrature formulas is realized and for the Volterra integral equations in the case when the external known function reduces to zero in the internal point of integration interval. Thus it is necessary to meet a coordination condition in the course of creation of the numerical solution.

This scheme of creation of the approximate solution of Volterra equation of the third kind can be distributed and for a case when the specified known function out of a sign of integration reduced to zero in several internal points of integration interval.

The received outcomes open a possibility of construction of a numerical solution of Volterra linear integral equations of the third kind with known function out of a sign of integration which is converted in zero in several points of a segment of integration. At certain cases the proved theorems can be spread and for the Volterra nonlinear integral equations of the third kind.

## References

1. Omurov, T.D., Karakeev, T.T. A Regularization and Numerical Methods of Solution of Inverse and No Local Boundary Value Problems. Bishkek (2006)
2. Magnitsky, N.A.: The Volterra linear integral equations of the first and third kind. Comput. Math. Math. Phys. **19**, 970–989 (1979)
3. Nakhushev, A.M.: The return tasks for the degenerating equations and the Volterra integral equations of the third kind. Differ. Equ. **10**, 100–111 (1974)
4. Karakeev, T.T., Rustamova, D.K., Bugubayeva, Zh.T.: Approximate Methods of the Solution of Volterra Linear Integral Equations of the Third Kind, pp. 6–10. Science and education, Prague (2014)
5. Karakeev, T.T., Rustamova, D.K.: Regularization and a Method of Quadrature for the Volterra Linear Integral Equations of the Third Kind, pp. 127–132. Studies on integral—different to the equations, Bishkek (2009)
6. Rustamova, D.K.: Regularisation of Volterra Linear Integral Equations of the Third Kind with Degenerated Function Out of an Integral in Two Points, pp. 52–57. Zh. Balasagyn Kyrgyz National University, Bishkek (2014)
7. Imanaliev, T.M., Karakeev, T.T., Omurov, T.D.: Regularization of the third-kind Volterra equations. Proc. Pak. Acad. Sci. **42**(1), 27–34 (2005)
8. Karakeev, T.T., Bugubaeva, ZH.T.: Equivalent of Transformation and a Regularization of Volterra Linear Integral Equations of the Third Kind, pp. 29–33. Zh. Balasagyn Kyrgyz National University, Bishkek (2012)
9. Glushak, A.V., Karakeev, T.T.: Numerical solution if the linear inverse problem for the Euler-Darboux equation. Comput. Math. Math. Phys. **46**, 848–857 (2006)
10. Karakeev, T.T.: A solution of Volterra integral equations of the third kind by method regularized quadrature. Bishkek Ser. **1**, 10–17 (2004)
11. Karakeev, T.T. Numerical Solution of Volterra Linear Integral Equations of the Third Kind, pp. 73–76. Samara State Technical University, Samara (2004)

# Cluster Analysis of Data with Reduced Dimensionality: An Empirical Study

**Pavel Krömer and Jan Platoš**

**Abstract** Cluster analysis is an important high-level data mining procedure that can be used to identify meaningful groups of objects within large data sets. Various dimension reduction methods are used to reduce the complexity of data before further processing. The lower-dimensional projections of original data sets can be seen as simplified models of the original data. In this paper, several clustering algorithms are used to process low-dimensional projections of complex data sets and compared with each other. The properties and quality of clustering obtained by each method is evaluated and their suitability to process reduced data sets is assessed.

**Keywords** Clustering · Metric multidimensional scaling · Sammon's projection · Affinity propagation · Mean shift · DBSCAN

## 1 Introduction

The majority of real-world phenomena such as audio and video signals, texts, fingerprints, and e.g. genome can be nowadays easily captured in the form of high-dimensional data. However, such data sets are often too large and too complex for efficient and accurate analysis using conventional methods and tools. Moreover, specific challenges such as presence of noise and curse of dimensionality often emerge and need to be dealt with [15].

Cluster analysis can be seen as partitioning of data into non-overlapping (hard) or overlapping (soft) classes consisting of objects with similar properties. The objects within a single cluster are supposed to be similar, i.e. the clusters are

P. Krömer (✉) · J. Platoš
IT4Innovations & Department of Computer Science, VŠB Technical University of Ostrava, Ostrava, Czech Republic
e-mail: pavel.kromer@vsb.cz

J. Platoš
e-mail: jan.platos@vsb.cz

© Springer International Publishing Switzerland 2016
V. Styskala et al. (eds.), *Intelligent Systems for Computer Modelling*,
Advances in Intelligent Systems and Computing 423,
DOI 10.1007/978-3-319-27644-1_12

121

supposed to be rather homogeneous, while the objects in different clusters are supposed to be dissimilar [8].

In this study, an experimental evaluation of the efficiency of several clustering algorithms applied to data with dimension reduced using two popular metric multidimensional scaling methods is provided. The rest of this paper is organized in the following way: Sect. 2 summarizes the investigated dimension reduction methods. Section 3 outlines the clustering algorithms used in this study and Sect. 4 details conducted computational experiments and their results. The work is concluded in Sect. 5.

## 2   Dimension Reduction

Dimension reduction is a general process of mapping data from an original, high-dimensional data space to a lower-dimensional feature (projection) space so that uninformative variance in the data is discarded and only a minimum amount of useful information is lost [4]. In this work, two traditional multidimensional scaling methods are used to model data in 2D on the basis of their distance matrices.

### 2.1   Classical Multidimensional Scaling

Classical metric multidimensional scaling (MDS) [1, 3, 14] is a method to transform a distance matrix into a set of coordinates in a $d$-dimensional space so that the distances between these coordinates are as close to the original distances as possible. The traditional (classical) MDS transforms the distance matrix into a cross-product matrix and finds its eigen-decomposition. During this process, the distance matrix is converted into a *similarity matrix* and its principal component analysis (PCA) is performed [1]. Alternatively, optimization methods that look for a configuration (i.e. lower-dimensional projection) $Y = (\mathbf{y}_1, \ldots, \mathbf{y}_m)$ that minimizes an arbitrary stress function [3, 11, 15], for example

$$S = \sum_{i<j}^{m} (d_{ij}^* - d_{ij})^2, \tag{1}$$

where $d_{ij}^*$ is the distance between a pair of the original data points, $\mathbf{x}_i$ and $\mathbf{x}_j$, and $d_{ij}$ is the distance between their corresponding lower-dimensional projections, $\mathbf{y}_i$ and $\mathbf{y}_j$, respectively.

## 2.2 Sammon's Projection

Sammon's projection (SP) is a non-linear metric multidimensional scaling method for projecting a data set from an original, high-dimensional data space to a space with lower dimensionality. It aims at preserving the between-point distances from the high-dimensional data space into the lower-dimensional projection space. This goal is achieved by minimizing an error criterion that penalizes the changes of distance between points in the original high-dimensional data space and in the low-dimensional projection space with an emphasis on preserving smaller distances [11].

Suppose that we have a collection $X$ with $m$ data points $X = (\mathbf{x}_1, \ldots, \mathbf{x}_m)$ where each data point $\mathbf{x}_i$ is an $n$ dimensional vector $\mathbf{x}_i = (x_{i1}, x_{i2}, \ldots, x_{in})$. At the same time we define a collection $Y$ of $m$ data points $Y = (\mathbf{y}_1, \ldots, \mathbf{y}_m)$ where each data point $\mathbf{y}_i$ is a $d$ dimensional vector and $d < m$. The initial values of the coordinates in $\mathbf{y}_i$ are chosen at random. The distance between vectors $\mathbf{x}_i$ and $\mathbf{x}_j$ is denoted $d_{ij}^*$ and the distance between corresponding vectors $\mathbf{y}_i$ and $\mathbf{y}_j$ in the lower-dimensional space is denoted $d_{ij}$. Here, any distance measure can be used to evaluate $d_{ij}$. However, the distance measure suggested originally by Sammon is the traditional Euclidean metric [13]. In that case, $d_{ij}^*$ and $d_{ij}$ are defined in the following way:

$$d_{ij}^* = \sqrt{\sum_{k=1}^{n}(x_{ik} - x_{jk})^2}, \quad d_{ij} = \sqrt{\sum_{k=1}^{d}(y_{ik} - y_{jk})^2}. \quad (2)$$

Projection error $E$ (so-called Sammon's stress) measures how well the current configuration of $m$ data points in the $d$-dimensional space matches the $m$ points in the original $n$-dimensional space

$$E = \frac{1}{\sum_{i<j}^{m}[d_{ij}^*]} \sum_{i<j}^{m} \frac{[d_{ij}^* - d_{ij}]^2}{d_{ij}^*}. \quad (3)$$

In order to minimize the projection error, $E$, any minimization technique can be used. Sammon's original paper from 1969 [13] used widely known methods such as pseudo-Newton (steepest descent) minimization

$$y_{ik}'(t+1) = y_{ik}'(t) - \alpha \frac{\frac{\partial E(t)}{\partial y_{ik}'(t)}}{\left| \frac{\partial^2 E(t)}{\partial y_{ik}'(t)^2} \right|}, \quad (4)$$

where $y_{ik}'$ is the coordinate of the data point's position $k$th $y_i'$ in the projected low-dimensional space. The constant $\alpha$ is usually taken from a range 0.3–0.4, originally proposed by Sammon. However, this range is not optimal for all types of problems. Equation (4) can cause a problem at the inflection points where second derivative is very small. Therefore the gradient descent may be used as an alternative minimization method.

# 3   Cluster Analysis

Clustering represents a fundamental data analysis task of separation of objects into meaningful clusters. A hard clustering of a data set $D = \{x_1, x_2, x_3, \ldots, x_n\}$ is a set $\mathcal{C} = \{C_1, C_2, \ldots, C_k\}$ composed of $k$ clusters $C_i$ subject to $C_i \subset D, C_i \neq \emptyset$ for each $C_i \in \mathcal{C}, \bigcup_{i=1}^{k} C_i = S$, and $C_i \cap C_j = \emptyset$ for each $C_i, C_j \in \mathcal{C}, i \neq j$.

There are different clustering methods that can be used for different types of data and for different kinds of data analysis analysis and processing (e.g. for object classification [2] vs. outlier detection). Most often used clustering algorithms include hierarchical clustering [8], centroid (medoid)-based clustering, and density-based clustering [12]. Unsupervised clustering of large data sets is a complicated NP-hard task. However, it is very attractive due to the countless applications in various fields of data science, machine learning, and e.g. data mining [2].

In this paper, three well-known clustering algorithms, representing different high-level clustering strategies, are used to find classes in data projected into a lower-dimensional space by MDS and SP, respectively.

## 3.1   Affinity Propagation

Affinity propagation (AP) is a clustering method based on passing messages between data points [9]. It aims at finding a set of high-quality *exemplars*, i.e. data points that represent classes of other points. They play a role similar to that of medoids in medoid-based clustering algorithms [8].

AP considers all points in the data set as potential exemplars and uses a heuristic procedure based on an iterative exchange of real-valued messages between data points to simultaneously select good exemplars and form corresponding clusters. It requires a similarity matrix, $\mathbf{S} = \{s_{ij}\}$, where $s_{ij}$ indicates how well the data point with index $j$ can serve as an exemplar of data point $j$, and iteratively updates two working matrices, responsibility matrix $\mathbf{R}$ and availability matrix $\mathbf{A}$ [9]:

$$\{r_{ij}\} = \{s_{ij} - \max_{k \neq j}(a_{ik} + s_{ik})\} \tag{5}$$

$$\{a_{ij}\} = \begin{cases} \left\{\min(0, r_{jj}) + \sum_{l, l \notin \{i,j\}} \max(0, r_{lj})\right\}, & i \neq j \\ \left\{\sum_{l \neq j} \max(0, r_{lj})\right\}, & i = j \end{cases} \tag{6}$$

The responsibility matrix reflects how good exemplar is $j$ to $i$ relative to other candidate exemplars and the availability matrix describes how proper is it for $i$ to pick $j$ as an exemplar with respect to preferences of other points in the data set. For each data point, $i$, the value of $j$ that maximizes $a_{ij} + r_{ij}$ labels $i$ as an exemplar (if $i = j$) or identifies the exemplar of $i$.

The algorithm is terminated after certain number of iterations, after the changes in the values of the messages fall bellow a predefined threshold, or after the exemplar decision (i.e. the clustering) remains unaltered for certain number of iterations. AP is summarized in Algorithm 1 [9].

**Algorithm 1: The affinity propagation algorithm**

1   $\{a_{ij}\} = \{r_{ij}\} = \{0\}$
2   **while** *Termination criteria not met* **do**
3      Update all responsibilities according to eq. (5)
4      Update all availabilities according to eq. (6)
5      Compute $a_{ij} + r_{ij}$ and perform clustering
6   **end**

## 3.2 Mean Shift Clustering

Mean shift, originally proposed in [10] and later popularized by [5], is a mode-based clustering method especially well-known in the field of computer vision. It does not require prior knowledge on the number of clusters or specific shape of clusters [6].

It is a *mode-seeking* algorithm that can be seen as a generalization of $k$-means like clustering algorithms. The algorithm is a kernel density estimator that iteratively seeks for the maxima of a density function represented by a set of discrete data samples. For a set of $n$ $d$-dimensional points, $X = \{\mathbf{x}_1, \ldots, \mathbf{x}_n\}$, radially symmetric kernel $K(x)$, and a bandwidth parameter (kernel radius) $h > 0$, the multivariate kernel density estimator $\hat{f}(x)$ is given by [6]

$$\hat{f}(x) = \frac{1}{nh^d} \sum_{i=1}^{n} K\left(\frac{\mathbf{x} - \mathbf{x}_i}{h}\right). \tag{7}$$

The *mean shift vector*, $\mathbf{m}_h(\mathbf{x})$, can be obtained from the gradient of $\hat{f}(x)$:

$$\mathbf{m}_h(\mathbf{x}) = \frac{\sum_{i=1}^{n} \mathbf{x}_i g\left(\left\|\frac{\mathbf{x}-\mathbf{x}_i}{h}\right\|^2\right)}{\sum_{i=1}^{n} g\left(\left\|\frac{\mathbf{x}-\mathbf{x}_i}{h}\right\|^2\right)} - \mathbf{x} \tag{8}$$

$$g(x) = -k(x) \tag{9}$$

where $k(x)$ is a kernel satisfying $K(\mathbf{x}) = c_{k,d} k(\|\mathbf{x}\|^2)$. The normalization constant $c_{k,d}$ is used to make sure that $K(\mathbf{x})$ integrates to 1. The mean shift vector, $\mathbf{m}_h(\mathbf{x})$, always points towards the direction of maximum increase in the density and defines integral curves that partition the data space [6]. An arbitrary point, $\mathbf{x}_i$, is assigned to a cluster, $C_j$, if the integral curve starting at $\mathbf{x}_i$ leads to mode $m_j$ of $\hat{f}(x)$. The mean shift clustering algorithm is summarized in Algorithm 2.

**Algorithm 2: The mean shift clustering algorithm**

1 **for** *each data point* $\boldsymbol{x} \in X$ **do**
2      **while** *not converged* **do**
3          Compute the mean shift vector $\boldsymbol{m}_h(\boldsymbol{x})$ according to eq. (8)
4          $\boldsymbol{x} = \boldsymbol{x} + \boldsymbol{m}_h(\boldsymbol{x})$
5      **end**
6      Assign $\boldsymbol{x}$ to the cluster corresponding to $\boldsymbol{m}_h(\boldsymbol{x})$
7 **end**

## 3.3  DBSCAN

The DBSCAN algorithm [2, 12] is a widely-used spatial density-based clustering method. Density-based clustering is useful due to its ability to discover clusters with arbitrary shapes but it suffers from high computational costs of cluster formation and clustering evaluation. Informally, a density based cluster $C_i$ is a set of points in the problem space that are *density connected*, i.e. for each pair of points in $C_i$ there is a chain of points with distance between two consecutive points smaller than a constant $\varepsilon$. Second parameter of the algorithm is the minimum number of points required to form a cluster, *minPts*. DBSCAN is outlined in Algorithm 3 [2].

**Algorithm 3: The DBSCAN algorithm**

1 Cluster index $i = 0$
2 **while** ***Not*** *All points have been labeled* **do**
3      Randomly select an unvisited *seed point*, $c_i$
4      Form the $\epsilon$-neighborhood of $c_i$, $\mathcal{N}(\epsilon, c_i)$. That is, find the set of all points that are *density connected* to $c_i$
5      **if** $|\mathcal{N}(\epsilon, c_i)| > minPts$ **then**
6          $C_i = \mathcal{N}(\epsilon, c_i)$
7          Expand $C_i$. That is, iteratively add to $C_i$ all unvisited points that are density connected to any point from $C_i$
8          $i = i + 1$
9      **else**
10          Mark $c_i$ as an outlier (noise)
11      **end**
12 **end**

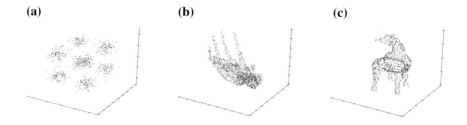

**(a)**                              **(b)**                              **(c)**

**Fig. 1** Test data sets in 3d. **a** Cube. **b** Hand. **c** Horse

## 4 Experiments

A series of computational experiments was performed in order to assess the ability
of different clustering methods to identify meaningful clusters in data sets with
dimensionality reduced by MDS and Sammon's projection. Two widely used and
one newly generated 3D data sets (point clouds) with clear structure and simple
cluster interpretation have been used in the experiments. The *Cube* data set was
prepared for this study. It consists of 800 points concentrated in clusters located in 8
corners of a 3-dimensional cube. The two remaining point clouds were taken from
the GeorgiaTech Large Geometric Models Archive[1] and down-sized by random
sampling to simplify the experimentation. The *Hand* data set contained a model of
human hand skeleton composed of 1636 points and the *Horse* data set models a
horse figure consisting from a total of 1212 points. A visualization the data sets is
shown in Fig. 1.

In our experiments, all three test data sets were projected to 2D by MDS and
Sammon's projection, respectively. Then, clustering by AP, MS, and DBSCAN
methods was applied to the 2D models of the original point clouds. The parameters
of each method were manually tuned on the *Cube* data set and remained fixed for
the remaining test cases. Each obtained clustering was visualized and assessed by
the Dunn index [7], a popular cluster validity measure

$$D = \min_{1 \le i \le n} \left\{ \min_{1 \le j \le n, i \ne j} \left\{ \frac{d(i,j)}{\max_{i \le k \le n} d'(k)} \right\} \right\}, \tag{10}$$

where $d(i, j)$ is the distance between clusters $i$ and $j$ and $d'(k)$ is the diameter of
cluster $k$. Different Dunn-like indexes use various definitions of $d(i, j)$ and $d'(k)$. We
have used the traditional $d(i, j)$ representing the shortest distance between any two
objects in $i$ and $j$ and $d'(k)$ expressing longest distance between any two points in
cluster $k$

---

[1]http://www.cc.gatech.edu/projects/large_models/.

**Table 1** Dunn index of clusterings obtained by every combination of *projection* and *clustering* methods

|       | MDS    |        |        | Sammon |        |        |
|-------|--------|--------|--------|--------|--------|--------|
|       | AP     | MS     | DBSCAN | AP     | MS     | DBSCAN |
| Cube  | 0.0173 | 0.0312 | 0.0052 | 0.0881 | 0.0881 | 0.0193 |
| Hand  | 0.0205 | 0.0068 | 0.0059 | 0.0124 | 0.0088 | 0.0098 |
| Horse | 0.0164 | 0.0072 | 0.0023 | 0.0231 | 0.0073 | 0.0056 |

A higher value of Dunn index indicates better clustering

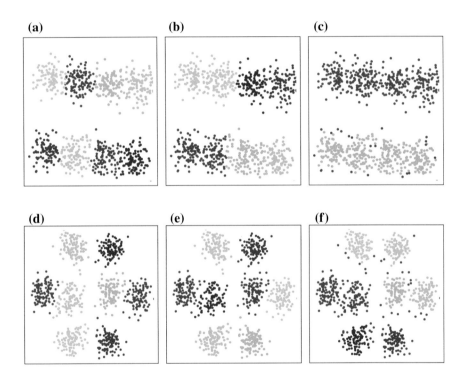

**Fig. 2** Visual results of experiments with the *Cube* data set. The caption of each subfigure indicates which *projection* and *clustering* method was used to generate it. **a** MDS and AP. **b** MDS and MS. **c** MDS and DBSCAN. **d** Sammon and AP. **e** Sammon and MS. **f** Sammon and DBSCAN

$$d(i,j) = \min_{a \in i, b \in j} \{dist(a,b)\}, \tag{11}$$

$$d'(k) = \max_{a,b \in k} \{dist(a,b)\}, \tag{12}$$

(a)                     (b)                     (c)

(d)                     (e)                     (f)

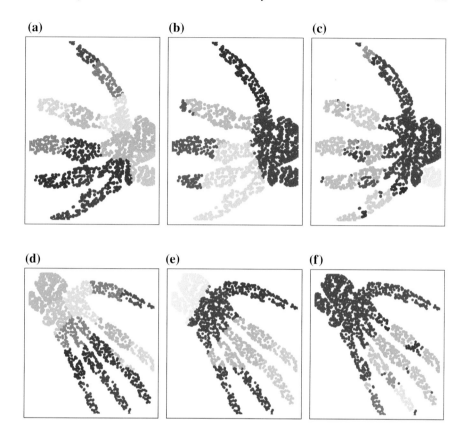

**Fig. 3** Visual results of experiments with the *Hand* data set. The caption of each subfigure indicates which *projection* and *clustering* method was used to generate it. **a** MDS and AP. **b** MDS and MS. **c** MDS and DBSCAN. **d** Sammon and AP. **e** Sammon and MS. **f** Sammon and DBSCAN

$$dist(a,b) = \sqrt{\sum_{i=1}^{n}(a_i - b_i)^2}. \qquad (13)$$

Results of performed experiments are summarized in Table 1 and visualized in Figs. 2, 3 and 4, respectively.

The results suggest that the quality of clustering depends on the type projection as well as on the properties of data. The Dunn index for clustering of data projected to 2D by Sammon's projection was higher (i.e. better) than that for clustering on data reduced by MDS for the *Cube* and *Horse* data sets regardless the clustering algorithm. For the *Cube* data set, AP and MS have obtained exactly the same clustering when Sammon's projection was used. The DBSCAN algorithm was, as the only one, able to identify outliers in the data. On the other hand, it was not able

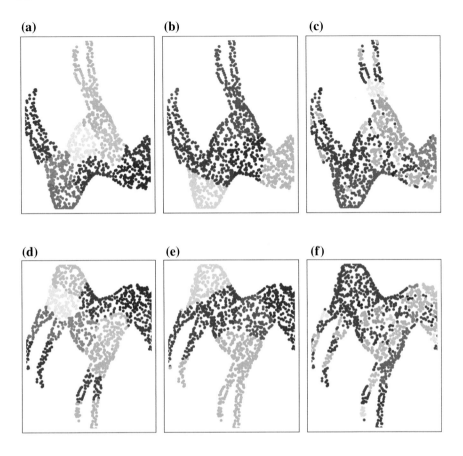

**Fig. 4** Visual results of experiments with the *Horse* data set. The caption of each subfigure indicates which *projection* & *clustering* method was used to generate it. **a** MDS and AP. **b** MDS and MS. **c** MDS and DBSCAN. **d** Sammon and AP. **e** Sammon and MS. **f** Sammon and DBSCAN

to distinguish between clusters when their points were projected close together, such as e.g. in Fig. 2c.

Different results were obtained for the *Hand* data set. In this case, the best clustering was obtained by a combination of MDS and AP. However, the Dunn index of clusterings obtained by MS and DBSCAN was better on 2D projections by the Sammon's mapping.

# 5 Conclusions

This study empirically investigated the influence of two different dimension reduction strategies, classic metric multidimensional scaling and Sammon's projection, on the clustering of various data set projected onto lower-dimensional feature spaces. 2D-projections of three data sets were clustered by means of affinity propagation, mean shift clustering, and DBSCAN algorithm, respectively. The resulting clusterings were assessed by the Dunn index. The Dunn index of the clusterings constructed by each method on data sets projected by Sammon's method was better than the Dunn index of the partitionings obtained on data sets mapped by the MDS method in 8 out of 9 test cases. That suggested that Sammon's mapping provides a more suitable projection and might be useful in cases when accurate parameters of the clustering procedure are unknown or cannot be tuned.

**Acknowledgements** This work was supported by the IT4Innovations Centre of Excellence project (CZ.1.05/1.1.00/02.0070), funded by the European Regional Development Fund and the national budget of the Czech Republic via the Research and Development for Innovations Operational Programme and by Project SP2015/146 of the Student Grant System, VŠB—Technical University of Ostrava.

# References

1. Abdi, H.: Metric multidimensional scaling. In: Salkind, N. (ed.) Encyclopedia of Measurement and Statistics, pp. 598–605. Sage, Thousand Oaks (2007)
2. Bandyopadhyay, S., Saha, S.: Unsupervised Classification: Similarity Measures, Classical and Metaheuristic Approaches, and Applications. SpringerLink, Bücher. Springer, Berlin (2012), https://books.google.cz/books?id=Vb21R9_rMNoC
3. Borg, I., Groenen, P., Mair, P.: Mds algorithms. In: Applied Multidimensional Scaling, pp. 81–86. SpringerBriefs in Statistics, Springer, Berlin (2013), http://dx.doi.org/10.1007/978-3-642-31848-1_8
4. Burges, C.J.C.: Dimension reduction: a guided tour. Found. Trends Mach. Learn. **2**(4) (2010), http://dx.doi.org/10.1561/2200000002
5. Cheng, Y.: Mean shift, mode seeking, and clustering. Pattern Anal. Mach. Intell. IEEE Trans. **17**(8), 790–799 (1995)
6. Comaniciu, D., Meer, P.: Mean shift: a robust approach toward feature space analysis. Pattern Anal. Mach. Intell., IEEE Trans. **24**(5), 603–619 (2002)
7. Dunn, J.C.: Well separated clusters and optimal fuzzy-partitions. J. Cybern. **4**, 95–104 (1974)
8. Everitt, B., Landau, S., Leese, M., Stahl, D.: Cluster Analysis. Wiley Series in Probability and Statistics, Wiley, Hoboken (2011), https://books.google.cz/books?id=w3bE1kqd-48C
9. Frey, B.J., Dueck, D.: Clustering by passing messages between data points. Science **315** (5814), 972–976 (2007), http://www.sciencemag.org/content/315/5814/972.abstract
10. Fukunaga, K., Hostetler, L.: The estimation of the gradient of a density function, with applications in pattern recognition. Inf. Theor. IEEE Trans. **21**(1), 32–40 (1975)
11. Hastie, T., Tibshirani, R., Friedman, J.: The Elements of Statistical Learning: Data Mining, Inference, and Prediction. Springer series in statistics, Springer, Berlin (2001), https://books.google.cz/books?id=VRzITwgNV2UC

12. Kriegel, H.P., Krüger, P., Sander, J., Zimek, A.: Density-based clustering. Wiley Interdisc. Rev. Data Min. Knowl. Disc. **1**(3), 231–240 (2011). doi:10.1002/widm.30
13. Sammon, J.W.: A nonlinear mapping for data structure analysis. IEEE Trans. Comput. **18**, 401–409 (1969)
14. Torgerson, W.S.: Multidimensional scaling: I. theory and method. Psychometrika **17**, 401–419 (1952)
15. Wang, J.: Geometric Structure of High-Dimensional Data and Dimensionality Reduction. Springer, Berlin (2012), https://books.google.cz/books?id=0RmZRb2fLpgC

# Comparison of Energy Near-Optimal Control Laws for the Drives with Constant and Linear Frictions

Ján Vittek, Branislav Ftorek, Peter Butko, Tomáš Fedor
and Ľuboš Struharňanský

**Abstract** Design and verification of two energy saving position control algorithms for drives loaded with combined constant and linear friction and load torques is main contribution of this paper. To decrease energy demands energy near-optimal control law based on copper losses minimization is compared with a control law based on symmetrical trapezoidal speed-time profile. Both algorithms respects prescribed maneuver time and have prescribed acceleration to achieve the demanded position. Overall control system exploits principles of vector control and forced dynamics control. A zero dynamic lag pre-compensator is included for precise tracking of prescribed state variables. Energy demands of both control algorithms are verified and compared by simulation, results of which confirmed possibility to achieve energy savings for pre-planned rest to rest maneuver.

## 1 Introduction

It is well-known that more than 50 % of generated electricity is consumed in motor drives [1]. Therefore even small decrease of drives' energy demands can result in significant energy savings and subsequent positive impact on environment. Efficient energy management of position controlled electric drives based on energy

J. Vittek (✉) · B. Ftorek · P. Butko · T. Fedor · Ľ. Struharňanský
Faculty of Electrical Engineering, University of Žilina, Žilina, Slovakia
e-mail: jan.vittek@fel.uniza.sk

B. Ftorek
e-mail: branislav.ftorek@fstroj.uniza.sk

P. Butko
e-mail: peter.butko@fel.uniza.sk

T. Fedor
e-mail: tomas.fedor@fel.uniza.sk

Ľ. Struharňanský
e-mail: lubos.struharnansky@fel.uniza.sk

© Springer International Publishing Switzerland 2016
V. Styskala et al. (eds.), *Intelligent Systems for Computer Modelling*,
Advances in Intelligent Systems and Computing 423,
DOI 10.1007/978-3-319-27644-1_13

133

near-optimal control for combined constant and linear friction or load torques is developed and verified as the main goal of this paper.

Two position control laws are developed and verified for comparison of their energy demands. The first one is based on conventional approach [2] to the energy minimization problem taking into account motor copper losses. Development of energy minimization control law is based on cost function respecting copper losses and Euler-Lagrange approach to minimization. The second approach is based on trapezoidal speed profile, having truly finite settling time and allowing precise prediction of all components of consumed energy [3].

Principles of *vector control* [4] and *forced dynamics control* (FDC) based on feedback linearization [5] are used to verify both energy saving control strategies.

Some progress on energy efficient position control of electric drives has already been made. The individual approaches to energy efficient position control of the drives differ mainly by description of velocity time profile [6]. Minimum energy motion control systems respecting principles of optimal control theory was developed by Sheta et al. [7] for the drive with dc motor. Similar approach was applied by Dodds et al. [8] for the drives with PMSM. Hamiltonian approach for generation of optimal trajectory of servo-drives was proposed by Wang et al. in [9]. A controller for position control of systems with energy optimization performances was proposed in [10].

## 2 Theoretical Background

To solve energy optimization task the state-space notation $\dot{x} = Ax + Bu$ and $y = Cx$ of the drive's dynamics is exploited. Control system is described with three differential equations for rotor position, $\theta_r$, speed, $\omega_r$, and electric torque, $\gamma_e$, which are completed with an equation for load torque, $\gamma_L = A + B\omega_r$ in which A and $B\omega_r$ respect Coulomb and viscous frictions respectively. Load torque derivative has form $\dot{\gamma}_L = B\dot{\omega}_r$.

$$
\begin{bmatrix} \dot{\theta}_r \\ \dot{\omega}_r \\ \dot{\gamma}_e \\ \dot{\gamma}_L \end{bmatrix} = \begin{bmatrix} 0 & 1 & 0 & 0 \\ 0 & 0 & 1/J_r & -1/J_r \\ 0 & -k_t^2/L_q & -1/T_s & 0 \\ 0 & 0 & B/J_r & -B/J_r \end{bmatrix} \begin{bmatrix} \theta_r \\ \omega_r \\ \gamma_e \\ \gamma_L \end{bmatrix} + \begin{bmatrix} 0 \\ 0 \\ k_t/L_q \\ 0 \end{bmatrix} u_q,
$$

$$
y = [1 \quad 0 \quad 0 \quad 0] \begin{bmatrix} \theta_r \\ \omega_r \\ \gamma_e \\ \gamma_L \end{bmatrix},
$$

(1a, b)

where $J_r$ is moment of inertia, $L_q$ and $R_s$ are stator inductance and resistance respectively, $T_s$ is stator time constant and $k_T = c\psi$ is motor torque constant, $u_q$ is excitation voltage of PMSM q_axis. Formulas for copper losses minimization is defined as:

$$I_0 = \frac{R_s}{k_T^2} \int_0^{T_m} \gamma_e^2 dt. \tag{2}$$

Complete cost function, $I_P$ for drive's copper losses minimization consists of minimization formula (2) completed with description of the state variables, into which Lagrange multiplication coefficients, $\lambda$ are added:

$$I_P = \frac{R_s}{k_T^2}\gamma_e^2 + \lambda_1\left(J_r\dot{\omega}_r - \gamma_e + \gamma_L\right) + \lambda_2\left(\dot{\theta}_r - \omega_r\right) + \lambda_3\left(\dot{\gamma}_L - B\dot{\omega}_r\right). \tag{3}$$

The solution of the problem requires calculation of Euler-Lagrange equations, found via partial derivatives of the designed cost function, $I_P$ for given state variables [11]:

$$\frac{\partial I_P}{\partial \theta_r} = \frac{d}{dt}\left(\frac{\partial I_P}{\partial \dot{\theta}_r}\right), \quad \dot{\lambda}_2 = 0. \tag{4a, b}$$

$$\frac{\partial I_P}{\partial \omega_r} = \frac{d}{dt}\left(\frac{\partial I_P}{\partial \dot{\omega}_r}\right), \quad -\lambda_2 = J_r\dot{\lambda}_1 - B\dot{\lambda}_3. \tag{5a, b}$$

$$\frac{\partial I_P}{\partial \gamma_e} = \frac{d}{dt}\left(\frac{\partial I_P}{\partial \dot{\gamma}_e}\right), \quad \frac{2R_s}{k_T^2}\gamma_e - \lambda_1 = 0, \tag{6a, b}$$

$$\frac{\partial I_P}{\partial \gamma_L} = \frac{d}{dt}\left(\frac{\partial I_P}{\partial \dot{\gamma}_L}\right), \quad \lambda_1 = \dot{\lambda}_3 \tag{7a, b}$$

where $\gamma_L = A + B\omega_r$ and A and $B\omega_r$ respect Coulomb and viscous frictions. Load torque derivative has form $\dot{\gamma}_L = B\dot{\omega}_r$.

Solving Euler-Lagrange Eqs. (5b) with regards (4b) and (7b) the general solution for electric torque can be written as:

$$\gamma_e = \lambda_1\frac{k_t^2}{2R_s} = \frac{k_t^2}{2R_s}\left[C_1e^{\frac{k_1}{J_r}} + \frac{c_1}{k_1}\right], \tag{8}$$

where c1 reflects (4b) and $C_1$ appears as solution of (5b). Calculation of both constants must respect boundary conditions for speed and position.

If into equation for rotor speed (1b) is inserted load torque expressed as $\gamma_L = A + B\omega_r$ then it has form:

$$\dot{\omega}_r(t) = \frac{1}{J_r}(\gamma_e - B\omega_r - A). \tag{9}$$

Integrating (9) and using constant variation method the formula of rotor speed results into form:

$$\omega_r = De^{-\frac{B}{J_r}t} + \frac{k_t^2}{2R_s}\left[\frac{C_1}{2B}e^{\frac{B}{J_r}t} + \frac{c_1}{B^2}\right] - \frac{A}{B}. \tag{10}$$

Taking into account boundary conditions for t = 0, $\omega_r$ = 0 and for t = $T_m$, $\omega_r$ = 0 results in calculation of constant D and equation for constants $C_1$ and $c_1$ as follows:

$$D = \frac{A}{B} - \frac{k_t^2}{2R_s}\left[\frac{C_1}{2B} + \frac{c_1}{B^2}\right], \tag{11a}$$

$$C_1\frac{k_t^2}{4R_sB}\left[e^{\frac{B}{J_r}T_m} - e^{-\frac{B}{J_r}T_m}\right] + c_1\frac{k_t^2}{2R_sB^2}\left[1 - e^{-\frac{B}{J_r}T_m}\right] = \frac{A}{B}\left[1 - e^{-\frac{B}{J_r}T_m}\right]. \tag{11b}$$

General formula for rotor angular speed has form:

$$\omega_r(t) = \left[\frac{A}{B} - \frac{k_t^2}{2R_s}\left(\frac{C_1}{2B} + \frac{c_1}{B^2}\right)\right]e^{-\frac{B}{J_r}t}$$
$$+ \frac{k_t^2}{2R_s}\left[\frac{C_1}{2B}e^{\frac{B}{J_r}t} + \frac{c_1}{B^2}\right] - \frac{A}{B}. \tag{12}$$

Finally if (12) is integrated then general function for rotor position, $\theta_r(t)$ is obtained as:

$$\theta_r(t) = \frac{-J_r}{B}\left[\frac{A}{B} - \frac{k_t^2}{2R_s}\left(\frac{C_1}{2B} + \frac{c_1}{B^2}\right)\right]e^{-\frac{B}{J_r}t}$$
$$+ \frac{k_t^2}{2R_s}\frac{C_1J_r}{2B^2}e^{\frac{B}{J_r}t} + \frac{k_t^2}{2R_s}\frac{c_1}{B^2}t - \frac{A}{B}t + D_1, \tag{13}$$

where $D_1$ is integration constant which can be determined via boundary conditions having form for t = 0, $\theta_r$ = 0 and for t = Tm, $\theta_r = \theta_{r\_dem}$ resulting in calculation of $D_1$ and relation for $C_1$ and $c_1$ as:

$$D_1 = \frac{J_r}{B}\left[\frac{A}{B} - \frac{k_t^2}{2R_s}\left(\frac{C_1}{2B} + \frac{c_1}{B^2}\right)\right] - \frac{k_t^2}{2R_s}\frac{C_1J_r}{2B^2} \tag{14a}$$

$$C_1\frac{J_r}{2R_s}\frac{k_t^2}{2B^2}\left[e^{\frac{B}{J_r}T_m} + e^{-\frac{B}{J_r}T_m} - 2\right] + c_1\frac{k_t^2}{2R_s}\left[\frac{J_r}{B^3}e^{-\frac{B}{J_r}T_m} + \frac{T_m}{B^2} - \frac{J_r}{B^3}\right]$$
$$= \theta_{r\ dem} - \frac{J_rA}{B^2}\left(1 - e^{-\frac{B}{J_r}T_m}\right) + \frac{A}{B}T_m \tag{14b}$$

If substitution Q = exp($k_1T_m/J_r$) is used then (11b) and (14b) arranged in matrix form enable to calculate constants $C_1$ and $c_1$ exploiting Cramer rule:

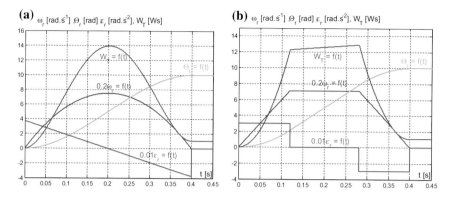

**Fig. 1** Energy near-optimal position control profiles of acceleration, speed and position including energy consumption. **a** Optimalized speed profile. **b** Trapezoidal speed profile

$$\frac{k_t^2}{2R_s}
\begin{bmatrix}
\frac{1}{2B}\left(Q-\frac{1}{Q}\right) & \frac{1}{B^2}\left(1-\frac{1}{Q}\right) \\
\frac{J_r}{2B^2}\left(Q+\frac{1}{Q}-2\right) & \left(\frac{J_r}{B^3Q}+\frac{T_m}{B^2}-\frac{J_r}{B^3}\right)
\end{bmatrix}
\cdot
\begin{bmatrix}
C_1 \\
c_1
\end{bmatrix}$$
$$=
\begin{bmatrix}
\frac{A}{B}\left(1-\frac{1}{Q}\right) \\
\theta_{r\,dem}-\frac{A}{B^2}J_r\left(1-\frac{1}{Q}\right)+\frac{A}{B}T_m
\end{bmatrix}. \tag{15}$$

Subplot (a) of Fig. 1 shows rotor position, speed, acceleration and energy consumption time profiles for the demanded position $\theta_{r\,dem} = 10$ rad and maneuver time, $T_m = 0.4$ s of the drive for energy near-optimal control respecting copper losses. Consumed energy for this maneuver was $W_T = 1.064$ Ws.

The second approach of energy near-optimal position control exploits symmetrical trapezoidal function for rotor speed together with related position profile approximating energy optimal variables quite closely. This strategy allows to reach demanded position via control on constant torque during acceleration and deceleration intervals and control on constant cruising speed in the middle interval. This approach was verified via analysis respecting Joule losses and friction losses together. Complete results for analysis of Coulomb, viscous and quadratic friction based on this approach was described in [12].

Magnitude of cruising speed, $\omega_p$ for trapezoidal speed profile is:

$$\omega_p = \int_0^{T_\varepsilon} \varepsilon_p dt = \varepsilon_p T_\varepsilon. \tag{16}$$

Demanded position, $\theta_{rd}$ as a function of acceleration, $\varepsilon_p$ and acceleration times, $T_m$, $T_\varepsilon$ respectively is defined as:

$$\theta_{rd} = 2 \int_0^{T_\varepsilon} \varepsilon_p t dt + \int_0^{T_0} \omega_p dt = \varepsilon_p \left( T_m T_\varepsilon - T_\varepsilon^2 \right). \tag{17}$$

Time of acceleration and deceleration, $T_\varepsilon$ are equal and valid under assumption:

$$T_\varepsilon = \frac{T_m}{2} \left[ 1 - \sqrt{1 - \frac{4\theta_{rd}}{\varepsilon_p T_m^2}} \right], \quad T_m \geq 2 \sqrt{\frac{\theta_{rd}}{\varepsilon_p}}. \tag{18a, b}$$

Constant cruising speed, $\omega_p$ is function of acceleration, $\varepsilon_p$ as:

$$\omega_p = \frac{T_m}{2} \left[ \varepsilon_p - \sqrt{\varepsilon_p^2 - \frac{4\theta_{rd}\varepsilon_p}{T_m^2}} \right]. \tag{19}$$

To check derived formula for motor losses and drive's total consumption the input energy is evaluated twice. Once as a sum of all drive's energies and secondly as an delivered input energy computed as:

$$W_d = \frac{3}{2} \int_0^{T_m} \left( u_q i_q + u_d i_d \right) dt. \tag{20}$$

The generation of required profiles respecting losses minimization is based on analysis of friction and copper losses of PMSM in steady state *(current dynamics due to inductances, $L_d$, $L_q$ and iron losses are neglected)* as shown in subplot (b) of Fig. 1.

For composed friction torque, $\Gamma_L = A + B\omega + C\omega^2$ description of predictable current is divided into three time intervals as:

$$i_{q1}(t) = I_{qd} + i_{qL}(t), i_{q2}(t) = I_{qL}, i_{q3}(t)$$
$$= - I_{qd} + i_{qL}(t), \tag{21}$$

where currents, $I_{qd}$ and $I_{qL}$ are constant:

$$I_{qd} = \varepsilon_p \frac{J_r}{c\Psi}, I_{qL} = \frac{\Gamma_{LT_0}}{c\Psi} = \frac{A}{c\Psi} + \frac{B\omega_p}{c\Psi} + \frac{C\omega_p^2}{c\Psi}, \tag{22}$$

Current, $i_{qL}$ is time varying current during acceleration periods:

$$i_{qL} = \frac{\Gamma_L(t)}{c\Psi} = \frac{A}{c\Psi} + \frac{B\omega}{c\Psi} + \frac{C\omega^2}{c\Psi}. \tag{23}$$

Following equation expresses total drive's losses, $W_T$:

$$W_T = \frac{3}{2} \int_0^{T_m} R_s i_q^2 dt + \int_0^{T_m} \left( A + B\omega + C\omega^2 \right) \omega dt. \tag{24}$$

Total losses are described as:

$$W_T = \frac{\theta_{rd}^2}{(T_m - T_\varepsilon)^2} \left[ \frac{k_1}{T_\varepsilon} + \frac{3R_s A^2}{2c\Psi^2} T_m + k_2 \left( T_m - \frac{4}{3} T_\varepsilon \right) + k_3 \frac{\theta_{rd}^2}{(T_m - T_\varepsilon)^2} \left( T_m - \frac{8}{5} T_\varepsilon \right) \right.$$
$$\left. + k_4 \frac{\theta_{rd}}{(T_m - T_\varepsilon)} \left( T_m - \frac{4}{3} T_\varepsilon \right) + k_5 \frac{\theta_{rd}}{(T_m - T_\varepsilon)} \left( T_m - \frac{3}{2} T_\varepsilon \right) + A\theta_{rd} \right], \tag{25a}$$

where:

$$k_1 = 3R_s \frac{J_r^2}{c\Psi^2}, k_2 = \frac{3}{2} R_s \frac{B^2}{c\Psi^2} + B, k_3 = \frac{3}{2} R_s \frac{C^2}{c\Psi^2},$$
$$k_4 = 3R_s \frac{AC}{c\Psi^2}, k_5 = 3R_s \frac{BC}{c\Psi^2} + C, \text{ and } k_s = k_2 + k_4. \tag{25b}$$

The optimized acceleration time, $T_\varepsilon$ for energy demands minimization is determine the derivative: $dW_T/dT_\varepsilon = 0$.

$$\frac{-4}{3} k_s T_\varepsilon^5 + \left[ \frac{10}{3} k_s T_m + 3k_5 \theta_{rd} \right] T_\varepsilon^4 + \left[ 3k_1 - \frac{2}{3} k_s T_m^2 - \frac{9}{2} k_5 \theta_{rd} T_m - \frac{24}{5} k_3 \theta_{rd}^2 \right] T_\varepsilon^3$$
$$+ \left[ -7k_1 + \frac{2}{3} k_s T_m^3 + \frac{3}{2} k_5 \theta_{rd} T_m^2 + \frac{12}{5} k_3 \theta_{rd}^2 T_m \right] T_\varepsilon^2 + 5k_1 T_m^2 T_\varepsilon - k_1 T_m^3 = 0, \tag{26}$$

which is the fifth order algebraic equation. To find zeros of (26) a Newton formula having quadratic convergence is used. The solution is optimized acceleration and deceleration time, $T_{\varepsilon \, opt}$.

$$x_{n+1} = x_n - \frac{f(x_n)}{f'(x_n)}, \quad n = 0, 1, 2 \ldots \tag{27}$$

Energy near-optimal acceleration time, $T_{\varepsilon \, opt}$ determines optimal acceleration, $\varepsilon_{p \, opt}$ as:

$$\varepsilon_{p \, opt} = \frac{\theta_{rd}}{T_{\varepsilon \, opt} \left( T_m - T_{\varepsilon \, opt} \right)}. \tag{28}$$

Subplot (b) of Fig. 1 shows rotor position, speed, acceleration and energy consumption time profiles for the demanded position $\theta_{r \, dem} = 10$ rad and maneuver time, $T_m = 0.4$ s of the drive for energy near-optimal control respecting copper and friction losses. Consumed energy for this maneuver was $W_T = 1.069$ Ws.

During investigation of energy consumption it was observed that for any speed dependent torque components (linear or quadratic) to decrease energy expenditures means to keep drive's cruising speed to the minimum practicable values. In opposite, for any combination of these two frictions with constant Coulomb friction brings only one unique solution of maneuver time, $T_m$.

To determine such maneuver time, $T_m$ for energy demands minimization the total energy (25a) was expressed in the terms of maneuver time as:

$$W_T = k_1 \frac{\theta_{rd}^2}{\kappa(1-\kappa)^2 T_m^3} + k_3 T_m + k_2 \frac{\theta_{rd}^2}{(1-\kappa)^2 T_m} \left(1 - \frac{4}{3}\kappa\right)$$
$$+ k_4 \frac{\theta_{rd}^4}{(1-\kappa)^4 T_m^3} \left(1 - \frac{8}{5}\kappa\right) + k_5 \frac{\theta_{rd}^3}{(1-\kappa)^3 T_m^2} \left(1 - \frac{3}{2}\kappa\right), \tag{29}$$

where coefficient, $\kappa$ is defined as a ratio between acceleration and maneuver time, $\kappa = T_\varepsilon / T_m$. Optimized maneuver time, $T_m$ is found exploiting relation: $dW_T/dT_m = 0$ as:

$$\frac{k_3(1-\chi)^2}{\theta_{rd}^2} T_m^4 - k_2 \left(1 - \frac{4}{3}\kappa\right) T_m^2 - \frac{2k_5\theta_{rd}}{(1-\kappa)} \left(1 - \frac{3}{2}\kappa\right) T_m$$
$$- \frac{3k_1}{\kappa} - \frac{3k_4\theta_{rd}^2}{(1-\kappa)^2} \left(1 - \frac{8}{5}\kappa\right) = 0. \tag{30}$$

The zeros of this function were found using Newton formula (27) with quadratic convergence speed.

Total energy consumption as a function of acceleration time $T_\varepsilon$ and maneuver time, $T_m$ is shown in Fig. 2 for friction coefficients for A = 1e − 2 Nm and B = 1e − 4 kg ms$^{-1}$.

**Fig. 2** Total energy consumption as a function of acceleration time, $T_\varepsilon$ and maneuver time, $T_m$

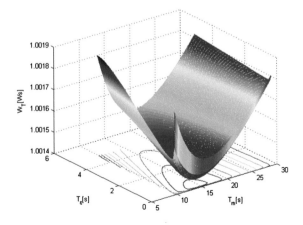

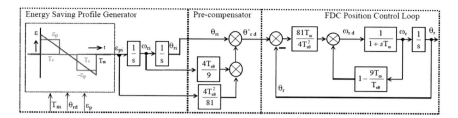

**Fig. 3** Block diagram for losses optimized position control of the PMSM

# 3 Position Control System Design

For verification of both proposed energy near-optimal position control algorithms the control structure capable of precise tracking of prescribed control variables was developed. Overall control system comprises a generator of energy saving acceleration, speed and position profiles completed with a zero dynamic lag pre-compensator and FDC based position control loop [13].

Inner speed control loop is based on FDC and respects also vector control principles [14]. Master control law is derived via feedback linearization, which for derivative of rotor speed, $d\omega_r/dt$ prescribes the first order dynamics, in which $T_\omega$ is the prescribed time and $\omega_d$ is the demanded rotor speed:

$$\frac{d\omega_r}{dt} = \frac{1}{T_\omega}(\omega_{rd} - \omega_r) \text{ or } \ddot{\theta}_r = \frac{1}{T_\omega}\left(\dot{\theta}_{rd} - \dot{\theta}_r\right). \qquad (31a,b)$$

Since the FDC speed control algorithm requires load torque information the observer with filtering capabilities [15] is implemented.

Speed control system was already described in details [16] and adjusting control system using Dodds formula was described [17].

Zero dynamic lag pre-compensator based on system inverse transfer function is exploited to achieve precise tracking of prescribed state variables. The overall energy saving position control system is shown in Fig. 3.

# 4 Verification of Designed Control System

Designed control algorithms for losses minimization were verified by simulation. Both algorithms were implemented for 17.8 kW PMSM parameters listed in the Appendix. To investigate energy expenditures the position control system of Fig. 3 prescribed state-variables for rest-to-rest maneuver.

Results of simulations are shown in Figs. 4 and 5 for moment of inertia of the motor only and Figs. 6 and 7 have enlarged moment of inertia 40times. For better comparison of the results gained the individual subplots show corresponding

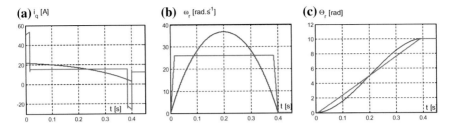

**Fig. 4** Time functions of stator current, load torque, rotor speed and position. **a** Stator current $i_q(t)$. **b** Rotor speed, $\omega_r$. **c** Rotor position $\theta_r$

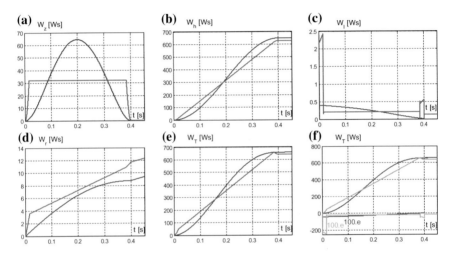

**Fig. 5** Losses components, mechanical energy and total delivered energy. **a** Rotational mass energy. **b** Shaft mechanical energy. **c** Magn. energy of inductances. **d** Energy covering copper losses. **e** Total energy. **f** Integrated input power

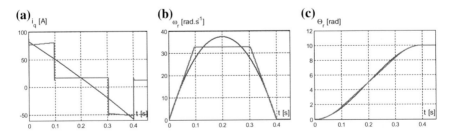

**Fig. 6** Time functions of stator current, load torque, rotor speed and position. **a** Stator current $i_q(t)$. **b** Rotor speed, $\omega_r$. **c** Rotor position, $\theta_r$

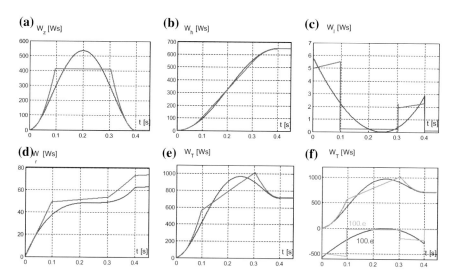

**Fig. 7** Energy consumption for drive with constant frictions and load torque. **a** Rotational mass energy. **b** Shaft mechanical energy. **c** Magn. energy of inductances. **d** Energy covering copper losses. **e** Total energy. **f** Integrated input power

state-variables for copper losses optimized law in 'blue' and for all losses optimized with trapezoidal speed profile in 'green'. Common requirements were: position demand, $\theta_{r\,d} = 10$ rad, prescribed maneuver time $T_m = 0.4$ s, and torque coefficients adjusted as: $A = 50$ Nm and $B = 0.5$ kg ms$^{-1}$.

The individual subplots of Figs. 4 and 6 show as subplot (a) stator current torque components where the influence of load torques changes is clearly visible. Load torque consisting of constant and speed dependent component are shown in subplot (b). Subplot (c) shows time functions of rotor speed and subplot (d) shows rotor positions time functions respectively.

Total energy expenditures including their complete break/down are shown in Fig. 5 for motor moment of inertia and Fig. 7 for 40 times enlarged moment of inertia respectively.

Subplot (a) of both figures time function of energy stored in mass of inertia. Subplot (b) shows delivered mechanical energy to cover shaft load. Energy stored in inductances is shown in subplot (c). Energy to cover copper losses is shown in subplot (d). Total energy consumption as a sum of listed components is shown in subplot (e). Finally subplot (f) shows total energy expenditures gained via integration of motor inputs (20) and compared with energy of subplot (e). Error between corresponding time functions is magnified 100×.

As can be seen from subplots (d) of Figs. 5 and 7 the copper losses of 'energy near-optimal' algorithm are in both cases lower if compared with algorithm based on trapezoidal speed profile. In opposite this isn't true for overall energy expenditures at low moment of inertia because the algorithm based on trapezoidal speed profile was already optimized for mechanical losses too. For enlarged moment of inertia due to increase copper losses the 'energy near-optimal' algorithm shows lower energy expenditures as expected.

# 5  Conclusions

Design and verification of the two energy minimizing control algorithms based on copper losses minimization was presented. Position control structure for PMSM exploited to evaluate energy expenditures covering load on the shaft and individual losses components and total energy expenditures operated proper way, which is confirmed by minimum error in two different ways of total energy expenditures evaluation.

As a further step of research is the optimization of the first proposed algorithm also for mechanical losses to achieve true 'energy-optimal control' and investigation of the derived 'energy optimal' and 'energy near-optimal' control algorithms experimentally.

**Acknowledgments**  This work was supported by R&D operational program Centre of excellence of power electronics systems and materials for their components I No. OPVaV-2008/2.1/01-SORO, ITMS 262220120003 and Centre of excellence of power electronics systems and materials for their components II. No. OPVaV-2009/2.1/02-SORO, ITMS 26220120046 funded by European regional development fund (ERDF). The authors also wish to thank the Slovak Grant Agency VEGA for support of research grant 1/0794/14.

# Appendix

PMSM parameters: $P_N$ = 17.8 kW, pole-pairs number p = 3, $R_s$ = 0.055 $\Omega$, phase inductance $L_q$ = 1.7 mH and $\psi_{PM}$ = 0.9 Vs, c = 3p/2. Nominal current $I_N$ = 32.4 A, total moment of inertial $J_r$ = 190.1 kg cm$^2$ and nominal torque $\Gamma_N$ = 85 Nm.

# References

1. Bose, B.: Energy, environment and advances in power electronics. IEEE Trans. Power Electron. **15**(4), 688–701 (2000)
2. Athans, M., Falb, P.L., Lacoss, R.T.: Time-, Fuel-, and energy optimal control of nonlinear norm-Invariant systems. IEEE Trans. Autom. Control **8**(3), 196–202 (1963)
3. Kim, C.H., Kim, B.K.: Energy saving 3-steps velocity control algorithm for battery-powered wheeled mobile robot. In: Proceedings of 2005 IEEE International Conference on Robotics and Automation, pp. 2375–2380. Barcelona, Spain (2005)
4. Novotny, D.W., Lipo, T.A.: Vector control and dynamics of AC drives. Clarendon Press, Oxford (1996)

5. Isidori, A.: Nonlinear Control Systems, 2nd edn. Springer, Berlin (1990)
6. Fedor, P., Perdukova, D.: Energy optimization of a dynamic system controller. AISC **189**, 361–369 (2013)
7. Sheta, M.A., Agarwal, V., Nataraj, P.S.V.: A new energy optimal control scheme for a separately excited dc motor based incremental motion. Int. J. Autom. Comput. 267–276 (2006)
8. Dodds, S.J.: Sliding mode vector control of PMSM drives with minimum energy position following. In: Proceedings of 2008 EPE-PEMC Conference, pp. 2559–2566. Poznan, Poland (2008)
9. Wang, Y., Ueda, K., Bortoff, S.A.: A Hamiltonian approach to compute an energy efficient trajectory for a servomotor system. Automatica **49**(12), 3550–3561 (2013)
10. Shyu, K.K., Lai, C.K., Tsai, Y.W., Yang, D.I.: A newly robust controller design for the position control of permanent-magnet synchronous motor. IEEE Trans. Industrial Electron. **49**(3), 558–565 (2002)
11. Manolea, G.: Loss-function optimal control of the positioning servomotors with static torque proportional to the speed. In: Proceedings of 7th International Workshop on Advanced Motion Control, AMC'02, pp. 232–235. Maribor, Slovenia, July 2002
12. Vittek, J., Bris, P., Pospisil, M., Butko, P., Fedor, T.: Energy saving position control of PMSM drives with constant, linear and quadratic torques. In: Proceedings of 2014 IEEE PEDES Conference, no. 70. Mumbai, India (2014)
13. Vittek, J., Dodds, S.J.: Forced dynamics control of electric drives. EDIS Publishing Centre of Zilina University, Slovakia (2003)
14. Boldea, I., Nasar, S.A.: Vector Control of AC Drives, 2nd edn. CRC Press, Boca Raton, FL (1992)
15. Brandstetter, P., Krecek, T.: Speed and current control of permanent magnet synchronous motor using IMC controllers. Adv. Electr. Comput. Eng. **12**(4), 3–10 (2012)
16. Vittek, J., Bris, P.: Energy saving position control algorithms for PMSM drives with Coulomb and viscous friction. In: Proceedings of IEEE ICCA conference, pp. 1485–1490. Hangzhou, China, June 2013
17. Dodds, S.J.: Settling time formulae for the design of control systems with linear closed loop dynamics. In: Proceedings of the International conference AC&T—Advances in Computing and Technology. University of East London, UK, 2007

# Systems of Temporal Logic for a Use of Engineering. Toward a More Practical Approach

**Krystian Jobczyk and Antoni Ligeza**

**Abstract** This paper is aimed at the evaluating of utility of 3 temporal logics: linear temporal logic (LTL), Allen's interval algebra and Halpern-Shoham interval logic from the point of view of the engineering practice. We intend to defend the thesis that chosen systems are only partially able to satisfy typical requirements of engineers.

## 1 Introduction

The practical utility of the commonly known temporal modal system: Linear Temporal Logic (LTL) or Intervals Allen's algebra ($All - 13$) seems to be indisputable, independently of their theoretical—or even philosophical provenance (see: [11]. In essence, temporal logic forms an essential component of many temporal planning tasks such as: generating of robot trajectories satisfying LTL formulas or temporal logic planning with using of model checking machinery. On the other hand, a precise characterization of the temporal systems ability in this matter (like their expressive power) is difficult—even from a 'purely' theoretical point of view. For example, it was shown by Maximova in [9] that LTL with operator 'next' does not respect the so-called Beth property, what means that not all implicit definitions of this system can be explicitly expressed in its language. This fact seems to justify that the expressive power of LTL is (at least partially) elusive. The similar difficulties with the expressive power evaluation stem from the more practical considerations concerning temporal systems. In fact, it is not clear, which properties of robot's activity and its work space in temporal logic motion planning can be still expressed in known temporal systems.

K. Jobczyk (✉)
University of Caen, Caen, France
e-mail: krystian_jobczyk@op.pl

K. Jobczyk · A. Ligeza
AGH University of Science and Technology of Kraków, Kraków, Poland

© Springer International Publishing Switzerland 2016
V. Styskala et al. (eds.), *Intelligent Systems for Computer Modelling*,
Advances in Intelligent Systems and Computing 423,
DOI 10.1007/978-3-319-27644-1_14

*Objectives of the paper*. With respect to it, we intend to propose a kind of retrospective evaluation of three well-known temporal systems from the point of view of their 'engineering' utility. We venture to formulate and defend a thesis that neither LTL, nor Allen's interval algebra, nor Halpern-Shoham logic are able to completely express the common engineering requirements imposed on (even typical) robot's activity in temporal logic based motion planning. For this purpose we distinguish a handful of such requirements concerning the system's specification, robot's actions and a nature of time. This issue forms a main purpose of this paper.

*Paper's motivation and state of art*. The main paper's motivation forms a lack of a broader comparative discussion on a real expressive power of well- known systems of temporal logic from the point of view of engineering requirements. The considerations of the paper stem from the earlier approaches to the time representation in the framework of LTL—introduced in [1977] by Amir Pnueli (a point-wise way), of the Allen's $All - 13$ algebra of 13 intervals relations—introduced by Allen in [1]— and of the Halpern-Shoham interval temporal logic introduced in [8]. These systems were widely discussed from the metalogical point of view in a seminal monograph [7]. The optimistic thesis about an utility of LTL in temporal logic motion planning for mobile robots was expressed in [5, 6] and in search control knowledge for planning in [3]. The role of LTL as a support of the discrete events based model was discussed by Antonniotti in [2]. A comparative monograph of Emerson [4] gives a broad overlook at the nature of the mutual relationships between modal logic and temporal logic. We are also interested in such temporal systems, which have a natural modal connotation or representation. This fact determined a subject of our analysis: LTL, Halpern- Shoham logic and the Allen's interval algebra. For the same purpose, we omit well-known Temporal Logic of Action of L. Lamport as a non-modal system, although actions in temporal framework will be a subject of our interest. It also appears that the evaluation of the Lamport's system requires some analysis, which essentially exceeds the thematic scope of this paper.

*Paper's organization*. In Sect. 2 we formulate our initial problem in a form of some paradigmatic example pf temporal planning with a robot per- forming tasks in block's world. The main paper's body forms Sect. 3, where we present three systems of temporal logic: LTL, Allen's interval algebra and Halpern-Shoham logic and we evaluate their ability to express the requirements imposed on the robot's task and its realization. In Sect. 4 we formulate concluding remarks and we give an outline of future research.

# 2   The Problem Formulation and Its Justification

It has been said that a main paper's objective is to evaluate the ability of chosen modal-temporal systems to express the several engineering requirements in examples of temporal planning. In order to realize this goal, we firstly extract a handful of such requirements from a paradigmatic example of the temporal planning with a robot performing the task to relocate the blocks in a given workspace

P. Secondly, we check which of the extracted engineering requirements (referring to the robot environments, robot tasks and their temporal requirements) can be captured by *Linear Temporal Logic*, the *Halpern-Shoham temporal logic* and *Allen's interval algebra*.

**Problem** We formulate the problem-example that will be addressed in this paper as follows: Consider a robot R that is able to move in a square environment with $k$-rooms $P_1, P_2, \ldots, P_k$ and a corridor *Corr* for some natural $k > 3$ with blocks A, B, C located somewhere in rooms $P_1, P_2, \ldots, P_k$. Consider that R performs the task: carry all the blocks and put them together in a corridor in an alphabetic order (firstly A, secondly B, finally C). Consider that the robot's activity has the following temporal constraints:

- Take a block B not earlier than $t_0 > 0$ after putting the block A in a corridor;
- Do not take two blocks in the same time;
- The room searching cannot be automatically finished by the robot;
- Visit the rooms $P_1, P_2, \ldots, P_k$ in any order;
- Since a moment $t_A$ visit the rooms in the order: $P_1, P_2, P_3$.

It easy to see that our problem seems to be a paradigmatic one for all class of similar problems and can be a convenient basis for further analysis and attempt of a new system construction. In fact, it contains typical commonly considered commands, tasks, actions concerning robot's activity and its admissible environment. Secondly, such a particularity degree corresponds well with a particularity degree of typical engineering requirements imposed on similar systems.

# 3 Engineering Requirements of the Problem-Situation

The above example allows us to distinguish the following engineering requirements imposed on the environment of the robot activity, its temporal constraints for its activity and the system specification.

**System specification**

1. *Sequencing*: Carry the blocks in alphabetical order: A, B, C.
2. *Coverage:* Go to rooms: $P_1, P_2, \ldots, P_k$.
3. *Conditions*: If you find a block A, B or C, take it; otherwise stay where you are.
4. *Conceptualization of the robot's activity*: Point-wise events (block A in in $P_k$ etc.) and actions on events as processes in time-intervals (the room searching by a robot R, the carrying of the blocks etc.)
5. *Nature of actions*: Some actions finish in the last action event, but some of them can last further automatically in a future (see: the room searching by a robot).

**Temporal requirements**

1. *Temporal sequencing*: Take a block B not earlier than $t_0 > 0$ after putting the block A in a corridor;
   Since the moment $t_A$ firstly visit the room $P_1$, after that $P_2$ and finally $P_3$.
2. *Temporal coverage*: Do not take two blocks in the same time;
3. *Action duration*: The duration time of some actions (like a room searching) can be longer than some time-interval $I_1$, but shorter than a time-interval $I_2$.
4. *Nature of time*: The states should be accessible from the 'earlier' states in a discrete linear time, but potentially—also in a continuous time.

## 3.1 Linear Temporal Logic (LTL) and Engineering Requirements

In order to evaluate whether LTL is able to express all of the desired engineering requirements, above extracted from the above problem-situation of the robot's activity, we will describe the syntax and semantics of LTL and distinguish the special class of LTL-formulas that could be especially useful for expressing of the above problem-situation.

*Syntax*. Bi-modal language of LTL is obtained from standard propositional language (with the Boolean constant T) by adding temporal-modal operators such as: *always in a past* (H), *always in a future* (G), *eventually in the past* (P), *eventually in the future* (F), *next and until* ($\mathcal{U}$) and *since* ($\mathcal{S}$)—co-definable with "until". The set *FOR* of LTL-formulas is given as follows:

$$\phi := \phi | \neg\phi | \phi \vee \psi | \phi\mathcal{U}\psi | \phi\mathcal{S}\psi | H\phi | P\phi | F\phi | Next(\phi) \qquad (1)$$

Some of the above operators of temporal-modal types are together co-definable as follows: $F\phi = T\mathcal{U}$, $P\phi = T\mathcal{S}\phi$ and classically: $F\phi = \neg G\phi$ and $P\phi = \neg H\phi$.

*Semantics*. LTL is traditionally interpreted in models based on the pointwise time-flow frames $\mathcal{F} = \langle T, < \rangle$ and dependently on a set of states $S$. In result, we consider pairs $(t, s)$ (for $t \in T$ representing a time point and $s \in S$) as states of LTL-models. Anyhow, we often consider a function $f : T \mapsto S$ that associates a time-point $t \in T$ with some state $s \in S$ and we deal with pairs $(t, f)$ instead of $(t, s)$. Hence the satisfaction relation $\models$ is defined as follows:

- $$(t,f) \models G\phi \Leftarrow (\forall t' > t)t' \models \phi, \quad (t,f) \models H\phi \Leftarrow (\forall t < t')t' \models \phi.$$

- $$(t,f)\models F\phi \iff (\exists t' > t)t' \models \phi, \quad (t,f)\models P\phi \iff (\exists t < t')t' \models \phi.$$

- $(t_1,f)\models \phi \mathcal{S}\psi \iff$ there is $t_2 < t_1$ such that $t_2, f \models \psi$ and $t, f \models \phi$ for all $t \in (t_1, t_2)$
- $(t_1,f)\models \phi \mathcal{U}\psi \iff$ there is $t_2 > t_1$ such that $t_2, f \models \psi$ and $t, f \models \phi$ for all $t \in (t_1, t_2)$

- $$(t_k,f)\models Next(\phi) \iff (t_{k+1},f)\models \phi, k \in \mathcal{N}\ominus.$$

***Specific set of formulas of LTL.*** Due to the observation from [6], we distinguish a class of special formulas of $\mathcal{L}(LTL)$ of two sorts. The first class $X = \{object -$
$-names : \psi_1^c, \psi_2^c, \ldots, P_1, \ldots, P_k, Corr, A, B, C; events : A^{P_1}, B^{P_2}, etc.\}$ will describe the robot's environments and its evolution; the second one—
$Y = \{actions : see(), move(), \ldots, go(), take(), a_1, a_2, \ldots etc.\}$—the robot's 'behaviour' and activity.

In accordance with our intentions, object-names will be denoted by concrete objects in a considered situations, the events-names by 'real' events such as that "block A is located in a room $P_1$ etc. In a similar way we encode actions as propositions. It not difficult to observe that LTL, enriched as above, is (at least partially) able to describe the situation of the robot's activity and partially express desired requirements as follows.

## System specification

1. *Sequencing*: Carry the blocks in alphabetical order: A, B, C.

$$F(Go(P_1)) \wedge F(Go(P_2)) \wedge \ldots F(Go(P_k)).$$

2. *Coverage*: Go to rooms: $P_1, P_2, \ldots, P_k$.

$$Go(P_1) \wedge Go(P_2) \wedge \ldots Go(P_k)$$

3. *Conditions*: If you find a block A, B or C, take it; otherwise stay where you are.

$$(See(A) \rightarrow take(A)) \wedge (See(B) \rightarrow take(B)) \wedge (See(C) \rightarrow take(C)).$$

## Robot's environments

2. The blocks A, B, C initially located somewhere in rooms $P_1, \ldots P_k$ but not in a corridor Corr.

$$A^{P_i} \wedge B^{P_j} \wedge C^{P_l} \wedge \neg(A^{Corr} \wedge B^{Corr} \wedge C^{Corr})$$

3. The corridor as a final place of the location of blocks A, B, C.

## Temporal requirements

2. Temporal coverage: Do not take two blocks in the same time:

$$\neg G(take(A) \wedge take(B) \wedge take(C))$$

3. *Nature of time*: The states should be accessible from the 'earlier' states in a discrete linear time, but potentially—also in a continuous time.

Independently of such a (relative big) expressive power of LTL, we can observe a difficulty with the expressing of such temporal requirements as delays and move of actions in time (after $t_o$, longer than $t_1$, but not shorten than $t_2$ etc.).

## 3.2 The Plan Construction and LTL

At the end of the paragraph we shall briefly evaluate how the LTL formalism can support plan constructions. We will refer to planning operators classically understood as a sequence of the appropriate actions (expressed in terms of its preconditions and effects). We will focus our attention on the robot's actions that we distinguished in the above engineering requirements. In our case we can approximate a plan construction as follows.

$go(r, P_1, P_2, \dots P_k)$ :
    robot $r$ goes to the room $P_1$ and-after that – to the adjacent rooms $P_2, \dots P_k$
    *preconditions:*  • adjacent$(P_1, P_2)$ ... adjacent $(P_{k-1}, P_k)$
                 • blocks are initially located somewhere in rooms $P_1, P_2 \dots P_k$,
                 but not in a corridor $Corr$:
                 $A^{P_i} \wedge B^{P_j} \wedge C^{P_l} \wedge \neq (A^{Corr} \wedge B^{Corr} \wedge C^{Corr})$
    *effects:*    $see(r, A^{P_i}) \wedge see(r, B^{P_j}) \wedge see(r, C^{P_l})$ for $i, j, l \in \{1, 2 \dots k\}$

$take(r, A, Corr)$
    robot r takes a block A from $P_i$ to the corridor $Corr$
    *preconditions:* • non-empty$(P_i)$, empty(Corr): $A^{P_j}, \neg Corr^A \wedge \neg Corr^B \wedge \neg Corr^C$
    *effects:* empty$(A^{P_i}$ ), non-empty(corridor):
    $\neg A^{P_i}, Corr^A$

$take(r, B, Corr)$
    robot r takes a block B from $P_j$ to the corridor $Corr$
    *preconditions:* • non-empty$(P_i)$, empty(Corr): $B^{P_j}, \neg Corr^A \wedge \neg Corr^B \wedge \neg Corr^C$
    *effects:* empty$(A^{P_i}$ ), non-empty(corridor):
    $\neg B^{P_i}, Corr^B$

$take(r, C, Corr)$
    robot r takes a block C from $P_j$ to the corridor $Corr$
    *preconditions:* • non-empty$(P_i)$, empty(Corr): $C^{P_j}, \neg Corr^A \wedge \neg Corr^B \wedge \neg Corr^C$
    *effects:* empty$(A^{P_i}$ ), non-empty(corridor):
    $\neg C^{P_i}, Corr^B$

## 3.3 Allen's 13-Algebra of Intervals

In the similar way, we intend to evaluate a utility of other temporal systems—introduced in [1] and commonly known as Allen's interval 13-algebra.

**Introduction + syntax.** This well-known temporal formalism is based on the observation that the relative position of each interval pair $(i, j)$ can be precisely done by one of the 13 bi-relations: $before(i, j)$, $meets(i, j)$, $overlaps(i, j)$, $during(t, j)$, $starts(z, j)$, $finishes(i, j)$ and their inverses (i.e., $before(j, i)$, $meets(j, i)$, etc.) and $equal(i, j)$. This formalism forms a kind of a (temporal) interval algebra with temporal interval $I$ defined by a condition:

$$\forall x, y \in I \forall z \in W (x < z < y \rightarrow z \in I) \tag{2}$$

The alphabet of $\mathcal{A}ll - 13$ contains of the above mentioned 13 binary predicates and interval variables $i, j$ (or eventually $k, l, \ldots$) and the Booleans. Formulas of $\mathcal{A}ll - 13$ is formed by Boolean combinations of the above listed atomic expressions.

**Semantics.** These 13 predicates are naturally interpreted as 13 relations between intervals. The semantics of Allen's algebra is presented on the following diagram.

**All-13 and the engineering requirements.** It easy to observe that $\mathcal{A}ll - 13$—defined as above—establishes only the mutual relations between temporal intervals and it is not enough to capture the situation of robot's activity in the blocks world. In order to make this system able to describe this situation, we enrich the language of $\mathcal{A}ll - 13$ by new basis-propositions: HOLDS(P, i) (property P holds during interval i) and OCCURS(E, i) (event E happens over interval i) and by propositions: $take()$, $go()$, $move()$ to denote the simple robot's actions. In particular, an event that a block A is located in $P_1$ over interval i can be expressed by OCCURS($A^{P_1}$). Such an extended system will be denote as $\mathcal{A}ll - 13^*$. In such a way, several of our desired engineering requirements can be easily expressed in $\mathcal{A}ll - 13^*$ (Fig. 1)

### System specification

1. *Sequencing:* Carry the blocks in alphabetical order: A, B, C.

   $HOLDS(take(A), i) \wedge HOLDS(take(B), j) \ldots \wedge HOLDS(take(C), k) \wedge starts(i, j)$
   $\wedge starts(j, k)$

2. *Coverage:* Go to rooms: $P_1, P_2, \ldots, P_k$.

   $HOLDS(go(P1)) \wedge HOLDS(go(P2)) \wedge HOLDS(go(P3))$

3. *Conditions:* If you find a block A, B or C, take it; otherwise stay where you are.

   $HOLDS(see(A), i) \rightarrow HOLDS(take(A, j)) \wedge meets(i; j)$

**Fig. 1** The visual
presentation of the temporal
relations of Allen

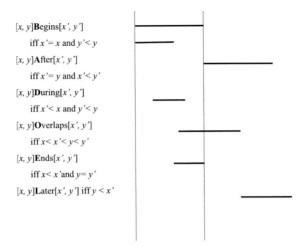

$[x, y]$**Begins**$[x', y']$

    iff $x' = x$ and $y' < y$

$[x, y]$**After**$[x', y']$

    iff $x' = y$ and $x' < y'$

$[x, y]$**During**$[x', y']$

    iff $x' < x$ and $y' < y$

$[x, y]$**Overlaps**$[x', y']$

    iff $x < x' < y < y'$

$[x, y]$**Ends**$[x', y']$

    iff $x < x'$ and $y = y'$

$[x, y]$**Later**$[x', y']$ iff $y < x'$

## Robot's environments

2. The blocks A, B, C initially located somewhere in rooms $P_1, \ldots, P_k$, but not in a corridor Corr.
3. The corridor as a final place of the location of blocks A, B, C.

The lack of direct possibility to express such adjectives as "initially" of "finally" can be avoided by referring the initial activities and robot's actions to some interval $i$ and final actions to some interval $j$ such that $i$ starts $j$ or—equivalently—$j$ finishes $i$. In this way we introduce, somehow, the chronology in the class of the considered events and actions. Then the conditions 2 and 3 (of robot's environment) can be formally represented as follows:

$$OCCURS(A^{P_1 \vee P_2 \ldots P_k}, i) \wedge OCCURS(B^{P_1 \vee P_2 \ldots P_k}, i) \wedge OCCURS(C^{P_1 \vee P_2 \ldots P_k}, i)$$
$$\wedge \neg OCCURS(A^{Corr}, j) \wedge OCCURS(B^{Corr}, j) \wedge OCCURS(C^{Corr}, j) \wedge finishes(j, i).$$

Independently of the relative expressive power of $\mathcal{A}ll - 13^*$, the possibility of this system still remain restricted. In essence, neither the temporal requirements and nature of time, nor the nature of actions (and their relations with the simple events) can be expressed in the propositional language of $\mathcal{A}ll - 13^*$.

*Remark* Note that an axiom, expressing time coverage ("Do not take blocks A, B, C in the same time"), of the form: $\neg \exists i (HOLDS(take(A), i) \wedge HOLDS(take(B), i) \wedge HOLDS(take(B), i)$ does not belong to the propositional $\mathcal{L}(\mathcal{A}ll - 13^*)$.

## 3.4 Shoham-Halpern Temporal Logic

This system (HS)—invented by Halpern and Shoham in [8]—forms a modified variant of earlier known modal logics of intervals, such as Moszkowski's pioneering interval based system from [10]. As earlier, we begin with a giving of its syntax and semantics to grasp an ability of this system to capture our engineering requirements.

**Definition** HS forms a modal representation of the following temporal relation between intervals, defined by Halpern and Shoham [8]: **a**fter" (or **m**eets"), (later"), **b**egins" (or start"), **d**uring", **e**nd" and **o**verlap". These relations are intuitive and their visualization can be easily found in many papers, so we omit their visual presentation, they correspond to the modal operators: $\langle A \rangle$ for **a**fter", $\langle B \rangle$ for **b**egins", $\langle D \rangle$ for **d**uring", etc. The syntax of HS entities $\phi$ is defined by:

$$\phi := p|\neg\phi|\phi \wedge \phi|\langle X \rangle|\langle \bar{X} \rangle, \tag{3}$$

where $p$ is a propositional variable and $\langle \bar{X} \rangle$ denotes a modal operator for the inverse relation wrt $X \in \{A, B, D, E, O, L\}$ If $\phi \in \mathcal{L}(\text{HS})$, $M$ is a model, and $I$ is an interval in the $M$-domain, then the satisfaction for the HS-operators looks as follows:

$$M, I \vDash \langle X \rangle \phi \Leftarrow \exists I' \text{ that } IXI' \text{ and } M, I' \vDash \phi. \tag{4}$$

**Halpern-Shoam logic and situation of robot's activity**. For the same purpose as in a case of $\mathcal{All} - 13$, we will enrich a language of HS by new propositions for an use of the describing of an environments and a 'behavior' of our robot. We extend the $\mathcal{L}(\text{HS})$ in the similar way as $\mathcal{L}(\mathcal{All} - 13)$. More precisely, we incorporate a handful atomic propositions as names of environments objects (rooms-names, blocks-names etc.), events and robot activities $(go, take, move, see)$ and two predicates of the form: $HOLDS(\phi_i^P)$ (property P holds during interval $i$) and $OCCURS(\phi_i^E)$ (event E happens over interval $i$).

In oder to measure an expressive power of **HS** with respect to Allen's algebra for a use of our situation of robot's activity, we introduce some translation $Tran : \mathcal{L}(\mathcal{All} - 13) \rightarrow \mathcal{L}(\text{HS})$ defined as follows:

$$tran(X(i,j)) = \langle X \rangle \phi_j, tran(OCCURS(E, i) = OCCURS(\phi_i^E),$$

$$tran(HOLDS(E, i)) = HOLDS(\phi_i^E),$$

where $X(i,j)$ is an Allen's relation between intervals and $\langle X \rangle \phi_j$ is the corresponding modal operator of $\mathcal{L}(\text{HS})$ for $X \in \{B, O, D, L, A, E\}$. Its immediate consequence is the following

**Fact** $\mathcal{All} - 13$ and HS* have the same expressive power wrt the temporal relation representation.

This fact allows us formulate a thesis that the same engineering requirements imposed on robot's activity in a 'block world'—expressed with a use of operators of $\mathcal{A}ll - 13$—can be alternatively expressed in a language of **HS\***. In particular, the sequencing condition: 'Carry the blocks in alphabetical order: A, B, C' can be therefore expressed in this way:

$$HOLDS(take(A)_i) \wedge HOLDS(take(B)_j) \wedge HOLDS(take(C)_k) \wedge \langle\ \rangle\phi_j \wedge \langle\ \rangle\phi_j.$$

In the same time, the same difficulties with the expressing of temporal requirements such as a linear and discrete nature of time are shared by **HS\***. It does not change, however, the fact that **HS**$^*$ is more 'sensitive' for a distinction between 'possible' and 'necessary' as-essentially-a modal system. The above established properties of considered temporal systems LTL, $\mathcal{A}ll - 13$ and **HS**$^*$ are presented in the table below.

| Properties | $\mathcal{L}$(LTL) | 13 Allen's algebra | HS* logic |
|---|---|---|---|
| • Time linearity | Yes | Move $r$ no | No |
| • Way of time representation | Pointwise | Intervals | Intervals |
| • Nature of time | Discrete, continuous | No | No |
| • Distinction between events and actions | Partial | Partial | Partial |
| • Different actions types | No | No | No |
| • Possibility to express the processes | Partial | Partial | Partial |
| • Possibility to express the moves and delays in time | No | No | No |
| • Representation of events in time | Non-concrete | Non-concrete | Non-concrete |

## 4  Conclusions and Future Works

In this paper, we have evaluated three important types of temporal modal logic: Linear Temporal Logic, Allen's interval algebra and Halpern-Shoham logic. We formulated and defended a thesis that these formalisms only partially satisfy typical engineer's requirements imposed on robot's activity in a block's world. We find this attempt promising for further extensions. The natural direction of the current investigation can be an evaluation of the expressive power of well-known temporal logic of action of L. Lamport—not only from the practical, but also from a theoretical point of view. It seems to be also promising to compare LTL with other powerful systems such as Transparent Intensional Logic (TIL).

Nevertheless, it appears that the most important common shortcoming of all these systems is their non-sensibility for different types of actions, processes and events. For example, these systems are not able to capture a difference between actions to events. Moreover, there is sometimes a need of a sharp distinguishing between actions that can last in a future independently of their initiator and its intentions and actions, which do not have such a property. This issue seems to be a promising subject of future research.

# References

1. Allen, J.: Maintaining knowledge about temporal intervals. In: Communications of ACM, vol. 26, no. 11, pp. 832–843 (1983)
2. Antonniotti, B., Mishra, M.: Discrete event models + temporal logic = supervisory controller: automatic synthesis of locomotion controllers. In: Proceedings of IEEE International Conference on Robotics and Automation (1999)
3. Bacchus, F., Kabanza, F.: Using temporal logic to express search control knowledge for planning. Artif. Intell. 116 (2000)
4. Emerson, A.: Temporal and modal logic. In: Handbook of Theoretical Computer Science, vol. B, pp. 995–1072 (1990)
5. Fainekos, G., Kress-gazit, H., Pappas, G.: Hybrid controllers for path planning: a temporal logic approach. In: Proceeding of the IEEE International Conference on Decision and Control, pp. 4885–4890. Sevilla, Dec 2005
6. Fainekos, G., Kress-gazit, H., Pappas, G.: Temporal logic moton planning for mobile robots. In: Proceeding of the IEEE International Conference on Robotics and Automaton, pp. 2032–2037 (2005)
7. Gabbay, D., Kurucz, A., Wolter, F., Zakharyaschev, M.: Many-Dimensional Modal Logics: Theory and Application. Elsevier (2003)
8. Halpern, J., Shoham, Y.: A propositional modal logic of time intervals. J. ACM **38**, 935–962 (1991)
9. Maximova, L.: Temporal logics with operator 'the next' do not have interpolation or beth property. In: Sibirskii Matematicheskii Zhurnal, vol. 32, no. 6, pp. 109–113 (1991)
10. Moszkowski, B.: Handbook of spatial logics. PhD-thesis, Stanford, Stanford University Press, 1983
11. van Benthem, J., Bezhanishvili, G.: Modal logic of space. In: Handbook of Spatial Logics, pp. 217–298. Springer (2007)

# Software Determining Optimum Order for Maintenance

**Vladimir Kral, Stanislav Rusek and Radomir Gono**

**Abstract** The paper deals with the development of Reliability Centered Maintenance software for distribution system network elements. In this case describes the order of maintenance for transformers 110 kV/MV. When determining an optimum maintenance sequence, the appropriate parameters of technical conditions and operational importance are required. The database of input values referring to the technical conditions and importance has been completed over the recent year. This paper describes processing of such input databases to create a format required for subsequent calculations and the first results that is the definition of order for maintenance.

**Keywords** 110 kV/MV transformer · Technical condition · Operational importance · Optimal order for maintenance · Database processing · RCM

## 1 Introduction

Definition of the optimal order for maintenance is based on the principle of Reliability Centered Maintenance (RCM). Its aim is to minimize the maintenance outage at particular devices while keeping their reliability and safety of the equipment in operation. RCM taking into account both the actual importance and condition of elements. Therefore, this maintenance is not based on time, as in case of preventive maintenance with each piece of equipment of the same kind treated

---

V. Kral · S. Rusek · R. Gono (✉)
Department of Electrical Power Engineering, VSB - Technical University of Ostrava,
Ostrava, Czech Republic
e-mail: radomir.gono@vsb.cz

V. Kral
e-mail: vladimir.kral@vsb.cz

S. Rusek
e-mail: stanislav.rusek@vsb.cz

© Springer International Publishing Switzerland 2016
V. Styskala et al. (eds.), *Intelligent Systems for Computer Modelling*,
Advances in Intelligent Systems and Computing 423,
DOI 10.1007/978-3-319-27644-1_15

equally regardless of its operational importance or technical condition. This maintenance process depends on the actual condition of the equipment.

We have been developing the methodology and software for maintenance focused on reliability by systematic means for 10 years at our Department of Electrical Power Engineering at VSB-Technical University of Ostrava. There were produced algorithms and calculation procedures for individual devices of the distribution system during this period. To be specific, methodology was developed for 110 kV power circuit breakers with $SF_6$, 110 kV power circuit breakers VMM type and 100 kV/MV transformers.

It is necessary to use real values of importance and technical condition in order to transfer results into practice. Whereas individual criteria, evaluation and weights of these criteria were continuously discussed with experts from the Distribution system operator (DSO). Thanks to their assistance we picked up the necessary values for the input database sourced from DSO technical information system and SAP. The data was limited to a certain number of these elements or a selected area. The main problem was always completeness and up-to-date status of the input databases.

## 2   Use of RCM for Condition Based Maintenance

The prerequisite necessary for definition of an optimal order for maintenance is the importance-condition relation. The relevant coefficients need to be used to specify importance and condition of particular elements and the values obtained shall be plotted into a 2D graph. The optimal order for maintenance of these elements is defined by their location in the 2D graph. Figure 1 represents a general

**Fig. 1** RCM principle by condition for 3 levels of maintenance activity

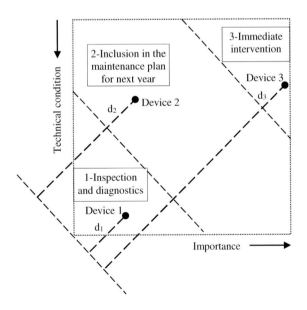

specimen graph, the order of maintenance (intervention) is determined by the length of segments $d_1 - d_3$. The higher the importance of equipment and the worse its technical condition, the sooner the maintenance should be launched.

Gradient of the line passing through the origin and 2nd and 4th quadrants for measurement of segment distance does not have to equal 45°. It is important to decide, whether the technical condition shall be preferred to operational importance, for example.

Apart from the technical condition, each element has to be identified too. Identification forms the third group of information about an element.

The data for every element within technical records (EWTR) is therefore divided into three groups:

- identification of a particular,
- data defining the EWTR status,
- data defining EWTR importance.

The procedure for optimization of maintenance by technical condition and importance, the details of methods for determination of technical condition and operational importance have been provided in [1], calculation of reliability characteristics for repairable units have been included in [2]. Reference [3] contains a clear definition for production of type transformers, including determination of limit values for maintenance priority.

General RCM principle as shown in the Fig. 1, where the result comprises 3 levels of maintenance activity, this applies in case of 2 values of operational importance only. However, the 110 kV/MV transformer produce the following 3 values—Transmitted energy, Transformer standby and Credit (Importance of connected MV lines). The details of all criteria of technical condition and importance, the permissible values and weight are also provided in [3]. The RCM principle for 4 levels of maintenance activity is shown in the Fig. 2.

# 3 Check Of The Data On Importance And Technical Condition

## 3.1 Sources and Range of Data

Our aim was to obtain all the input values necessary for calculation using the software for definition of maintenance sequence by means of our templates completed by personnel from the Distribution system service company. However, the individual values were hard to obtain by complex means in terms of experts from the relevant information systems. Some values had to be obtained manually by additional searches, some values were impossible to obtain (referring to the technical condition at Number of overload hours, Short-circuit intensity on the lower voltage side—that was why these criteria were excluded from the assessment of technical condition), others had to be finished by calculation using alternate values (referring

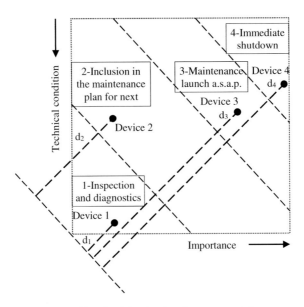

**Fig. 2** RCM principle by condition for 4 levels of maintenance activity

to the importance of Transmitted energy and Credit). Moreover, the personnel from the Distribution system service company had to compile a methodology for collection of such data to eradicate differences among results for individual areas.

The data for creation of input databases required for processing in the calculation software to define the order of maintenance was finally obtained from "two sources". The input values referring to the technical condition of transformers were provided by the personnel from the Distribution system service company and the importance values came from the personnel of the Distribution system operator.

The double source of data led to the situation with slight differences in identification of transformers in the databases of technical conditions and importance respectively. Further consequences include the fact that some transformers were missing either from the database of technical condition or the importance database.

The data on technical condition was supplied in form of a separate table per area, where the assessment of particular criteria complied with the markings stated in the data collection methodology.

The importance data had to be assembled from three separate sources per area.

The first file with "compliant" identification of transformers as the technical condition data file contained the values referring to transformer (TR) standby and high voltage (HV) lines standby required for definition of the criterion for the resultant TR standby. These two values were further supplied with additional two constant values for all transformers pursuant to agreement with the client—Rise time of TR (5 min) and the Rise time for HV line (120 min).

Another file contained the details of consumers connected per category. These values were used to calculate another criterion of importance—Credit (Importance of MV lines connected). The last group of files (different quantities per area) contained the exports of annual energy flow for the calendar year of 2013. These were processed to calculate the Energy transmitted and the Operation period (auxiliary detail). The two values were further used to calculate the criterion of technical condition for Average hourly load.

## 3.2 Required Versus Supplied Data

The Table 1 shows a specimen of descriptive data already processed for transformers corresponding with our template, the Table 2 contains the descriptive data supplied.

The Table 3 shows a specimen of values referring to the technical condition in our template, the Table 4 shows the values supplied.

The templates were created using the structure of data from the input values supplied in the past. The tables above clearly show that the records for identification of individual transformers in the information systems have changed. Assessments of individual criteria outside the specific date were admissible within the range of 0–100, the internal methodology of data collection defined the interval of 1–X, where X refers to the number of potential conditions.

**Table 1** Descriptive details of transformers required

| Area | Abbreviation | Transformer | Year of manufacture | Serial number |
|------|--------------|-------------|---------------------|---------------|
| 1 | D1_Ss01 | T101 | 1998 | 154408 |
| 1 | D1_Ss01 | T102 | 1999 | 154409 |
| 1 | D1_Ss02 | T101 | 1975 | 0941 026 |

**Table 2** Descriptive details of transformers supplied

| Area | ID station | District | Serial number | Year of manuf. |
|------|------------|----------|---------------|----------------|
| 1 | D1_Ss01//OST01/T101 | D1 | 154408 | 1998 |
| 1 | D1_Ss01//OST02/T102 | D1 | 154409 | 1999 |
| 1 | D1_Ss02//OST01/T101 | D1 | 0941 026 | 1975 |

**Table 3** Format of the technical condition values required

| Weather conditions | Control type | Year of manufacture of the machine | Transformer tank condition | Leaks |
|--------------------|--------------|-----------------------------------|----------------------------|-------|
| 100 | 100 | 1998 | 100 | 100 |
| 100 | 100 | 1999 | 100 | 100 |
| 100 | 100 | 1975 | 100 | 75 |

**Table 4** Format of the technical condition values supplied

| Weather conditions | Control type | Year of manufacture of the machine | Transformer tank condition | Leaks |
|---|---|---|---|---|
| 1 | 1 | 1998 | 1 | 1 |
| 1 | 1 | 1999 | 1 | 1 |
| 1 | 1 | 1975 | 1 | 2 |

**Table 5** Importance values required

| Energy transmitted (GWh) | TR standby (%) | HV lines standby (%) | Rise time of TR (min) | Rise time of HV line (min) | Credit (-) |
|---|---|---|---|---|---|
| 75.02 | 100 | 90 | 5 | 120 | 30681 |
| 111.29 | 100 | 90 | 5 | 120 | 30681 |
| 59.00 | 100 | 100 | 5 | 120 | 22990 |

**Table 6** Supplemental values for Credit calculation

| Credit | Number of RSs connected | Number of CSs connected | Number of WCs connected | Hours of operation |
|---|---|---|---|---|
| 30681 | 15991 | 2148 | 79 | 8704 |
| 30681 | 15991 | 2148 | 79 | 8717 |
| 22990 | 13005 | 1597 | 40 | 7337 |

Table 5 contains the importance values required. Personnel from the Distribution system operator supplied the values for TR standby and HV line standby only.

Calculation of Credit values was aided by lists of distribution transformers showing particular consumer categories. The values of Energy transmitted were calculated using details from files that contain annual measurements of energy flows. That was the reason to create the auxiliary Table 6 that shows supplemental values for Credit calculation (RS—residential sector, CS—commercial sector, WC —wholesale consumers).

The importance data came from another source, where the ID station details would also differ in some cases, despite their association with the same transformer. Those 2 input databases then had to be paired in a certain way.

The logically simplest way was to pair individual transformers according to the ID station detail or the transformer serial numbers. The ID station detail contains the district (D) abbreviation, the very 4-character code of substation (Ss) and the number of transformer. The best match obtained after testing at every area resulted from the search using the first 7 characters—the district and the substation and 4 last characters of the ID station—transformer numbers/stands (Table 2). The more logical pairing by serial numbers was less successful due to inconsistent records of these numbers in the input tables—combination of the text/number format, recording of insignificant 0 digits at the beginning, spaces etc.

The database completeness check was conducted in two steps. The first one involved a cross check between technical condition and importance and vice versa with exclusion of all transformers showing no matches found. The lists of those were generated into relevant files per area. The second step excluded those transformers not matched by lists of consumers connected or the relevant files with details on energy flows.

The values of all criteria (technical condition and importance) per area were inserted into independent Excel files. These full details were then used to generate all the input databases required by the calculation software for determination of sequence of transformers for maintenance purposes.

# 4  Input Data Adaptation for the Software Requisites

## 4.1  Data Adaptation for the Software According to Primary Requirements

The diversity in assessment of technical condition criteria defined by the methodology applied for their collection required a conversion compliant with the template for calculation software. That had to be paralleled by pre-processing of the details on technical condition as well as importance in every area to ensure mutual compliance of their structure. The number of descriptive details and sequences of particular criteria was actually different in tables addressing individual areas.

## 4.2  Forced Data Adaptation According to Real Databases

The second stage addressed the requirement to remove ambiguities at every criterion, anomalies, different descriptions for the same item (e.g. the values at Number of switch-overs did not include any counters or were listed as undefined), potential typing errors, double details or incorrect format of dates (text details, date not expressed in numbers only). Anomalies occurring to a minor extent were corrected manually, the remaining ones were left for software correction.

## 4.3  Software Adaptations

The very calculation software for determination of the maintenance sequence was merely subject to alterations associated with the change of certain descriptive details on transformers and their maximum permissible length. However, the issues to be tackled concerned processing of the necessary input databases for this software using the data supplied, together with calculation of Credit, Energy transmitted and the Average hourly load. The VBA option in Excel was selected to prevent any interventions with the calculation algorithm of this software.

We created 4 supporting programs to process the databases in a gradual process and generate the input databases required for the very calculation software. The first one serves for conversion of the technical condition values supplied with respect to methodology of their collection into an appropriate format (see Tables 3 and 4), the second one will assign these details to values of importance criteria values and perform the cross check of completeness. The third file with the number of consumers per category will be used for calculation of Credit; the last one will use the energy flows measured to calculate the total Energy transmitted and the Operation period to make an additional calculation and supplement the Average hourly load criterion.

## 5 Maintenance Order Evaluation for 110 kV/MV Transformers

The calculation algorithm, the description of calculation program and the associated operation procedure are described in detail by [3] and they match the flow chart in Fig. 3.

The final step involved calculation of maintenance sequence and determination of the relevant maintenance activity per specific area for all transformers provided with full input details on technical condition and importance. The numbers of transformers per group arranged by maintenance activities are shown in Table 7. The Table 8 then shows the maintenance sequence for one of the areas; that is for the first 10 and the last 5 transformers respectively.

The main reason why a particular transformer falls within the maintenance groups 2 or 3 is its technical condition, especially the criterion with greatest weight— Diagnostics, as well as the criteria associated with the age of transformer and the date of routine maintenance in combination with the high value of importance.

**Fig. 3** Block
diagram of the program

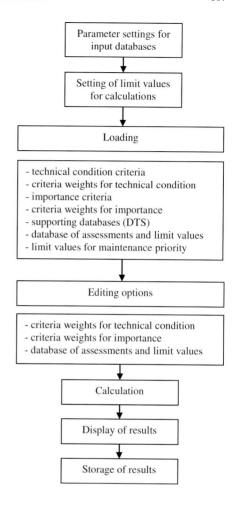

**Table 7** Numbers of transformers in individual groups per maintenance activity

| Area | Transformers | Number in group per maintenance activity | | |
|---|---|---|---|---|
| | Total | 1 | 2 | 3 |
| 1 | 71 | 67 | 4 | 0 |
| 2 | 56 | 48 | 8 | 0 |
| 3 | 76 | 69 | 7 | 0 |
| 4 | 50 | 43 | 5 | 2 |
| 5 | 89 | 78 | 9 | 2 |
| Total | 342 | 305 | 33 | 4 |

**Table 8** Maintenance sequence per area 1

| Abbreviation | Transformer | Year of manufacture | Serial number | Apparent power (MVA) | Technical condition (%) | Diagnostics (%) | Importance (%) | Priority (%) | Sequence | Threshold limit |
|---|---|---|---|---|---|---|---|---|---|---|
| D3_Ss22 | T101 | 1990 | 0962 528 | 40 | 29.8 | 100 | 60.0 | 65.1 | 1 | 2 |
| D3_Ss22 | T102 | 1993 | 0965 908 | 43 | 23.9 | 100 | 54.0 | 65.1 | 2 | 2 |
| D8_Ss16 | T101 | 1972 | 938023 | 25 | 29.7 | 100 | 54.0 | 62.1 | 3 | 2 |
| D8_Ss16 | T103 | 1987 | 956415 | 40 | 43.2 | 100 | 60.0 | 58.4 | 4 | 2 |
| D4_Ss13 | T101 | 2009 | 968795 | 40 | 40.6 | 100 | 54.0 | 56.7 | 5 | 1 |
| D6_Ss08 | T103 | 1982 | 0944 589 | 40 | 48.7 | 100 | 60.0 | 55.6 | 6 | 1 |
| D6_Ss08 | T101 | 1998 | 929802 | 25 | 39.3 | 100 | 48.0 | 54.4 | 7 | 1 |
| D6_Ss08 | T102 | 1967 | 929801 | 40 | 48.4 | 100 | 54.0 | 52.8 | 8 | 1 |
| D9_Ss04 | T102 | 1994 | 921229 | 40 | 39.8 | 66 | 39.8 | 50.0 | 9 | 1 |
| D2_Ss07 | T102 | 2005 | 968183 | 40 | 49.7 | 100 | 49.7 | 50.0 | 10 | 1 |
| ... | ... | ... | ... | ... | ... | ... | ... | ... | ... | ... |
| D1_Ss03 | T102 | 2008 | 462083 | 40 | 87.2 | 100 | 40.6 | 26.7 | 66 | 1 |
| D4_Ss12 | T103 | 2000 | 932922 | 25 | 70.6 | 100 | 21.1 | 25.3 | 67 | 1 |
| D4_Ss11 | T102 | 1988 | 958404 | 25 | 73.4 | 100 | 20.4 | 23.5 | 68 | 1 |
| D7_Ss20 | T102 | 1993 | 932921 | 25 | 74.8 | 100 | 19.7 | 22.5 | 69 | 1 |
| D7_Ss09 | T102 | 1997 | 926334 | 25 | 81.5 | 100 | 22.2 | 20.4 | 70 | 1 |

# 6 Conclusion

This article deals with results of task from industry to specify optimal maintenance order of 110 kV/MV transformers. That was processing of an actual database of transformers with the first results. Results are partial only because not all transformer has been supplied with all the required input data yet. Once these transformers have been completed, we will have a full database of all 110 kV/MV transformers managed by the Distribution system operator available.

The initial summarized results listed in the Table 7 show that no transformer falls within the category 4 for immediate shutdown and there are just 4 units to be maintained as soon as possible. There is almost 90 % of transformers falling within the level for maintenance activity 1—inspection and diagnostics.

**Acknowledgments** This research was partially supported by the SGS grant from VSB - Technical University of Ostrava (No. SP2016/95) and by the project TUCENET (No. LO1404).

# References

1. Mourbay, J.: Reliability-Centred Maintenance. Butterworth-Heinemann, Oxford (1997)
2. Martínek, Z., Královacová, V.: The solution for repairable units. In: Proceedings of the 11th International Scientific Conference Electric Power Engineering 2010, University of Technology Brno, pp. 593–597 (2010)
3. Král, V., Rusek, S., Goňo, R.: Software for the assessment of the optimum maintenance sequence for 110 kV/MV transformers. In: Proceedings of Elnet 2010, VŠB-TU Ostrava, Ostrava, pp. 15–26 (2010)

# Statistical Evaluation of Dimmable Interior Lighting System Consumption Using Daylight

Tomáš Novák, Petr Bos, Jan Šumpich and Karel Sokanský

**Abstract** From the power engineers point of view, at the present time, the necessity to predict the consumption of lighting systems becomes more and more urgent at the time of the project phase. The reason is to compare the projects from different investors and especially to compare the projects with regulated (dimmed) lighting systems (these are regulated at a constant level of illumination) with the projects of unregulated lighting. In that case, the designer has to be able to determine the payback of regulated lighting system in comparison with classic unregulated lighting systems. Nowadays, it is possible to calculate the potential savings. The calculating is determined by the LENI (Lighting Energy Numeric Indicator), according to EN 15193, which is based upon the analytical estimate of potential savings calculated on the base of empirically derived coefficients for standardized rooms, standardized operations, etc. However, this model is not based on real specific situations. Therefore, the team of VSB-TU Ostrava developed a predictive model, which is based on the daylight dynamics in specific lighting conditions. The long-term measurements were carried out to verify whether the model´s results correspond to the real facts. The statistical evaluation of the model´s conclusions and the real measured rates was carried out as well.

**Keywords** Predictive model · Regulated lighting systems · Constant level of illumination · Daylight · Energy savings

## 1 Introduction

The proposed prediction model uses an artificial interior lighting system which can be dimmed to a constant illuminance level and deals with the contribution of daylight entering the space through windows and skylights. This developed model

T. Novák (✉) · P. Bos · J. Šumpich · K. Sokanský
Faculty of Electrical Engineering and Computer Science, VŠB—Technical University of Ostrava, Ostrava, Czech Republic
e-mail: tomas.novak1@vsb.cz

© Springer International Publishing Switzerland 2016
V. Styskala et al. (eds.), *Intelligent Systems for Computer Modelling*,
Advances in Intelligent Systems and Computing 423,
DOI 10.1007/978-3-319-27644-1_16

gives transparent and unambiguous evidence of the economic benefit of using lighting systems that reduce the intensity of light emitted by them depending on the amount of daylight entering the rooms that are being illuminated.

## 2   Proposal of the Prediction Model

Diffuse illumination is being increasingly used in the interior lighting systems, a combination of daylight with artificial light. The technical level of the current lighting systems enables the regulation of the luminous flux in lighting control systems at a constant level of illumination due to the artificial lighting dimming systems of each lamp. Using daylight in these lighting systems can save a considerable part of electricity.

Based on the technical design of the interior lighting system and cyclic calculations of the dimming levels for light lines, or even for the individual lights, the dimming development patterns are obtained for the light lines or individual lights.

The dimming figures are converted to electric power enabling the power used for the lighting system to be calculated depending on the uniformly overcast sky in any given time.

The base for dynamic modeling of daylight is uniformly overcast sky, as it is mentioned above. Due to that the influence of windows positions towards the cardinal points can be eliminated. In the calculation we do not consider only the uniformly overcast sky, but also the influence of the Sun declination that is changing throughout the days and year. The external intensity of illumination varies not only during the day, but also during the whole year. Value of illuminance is also dependent on the time and the clock angle of the Sun, on the Solar elevation and on the diffuse illuminance [1].

### 2.1   The Daylight Factor

The daylight factor expresses the daylight illumination contribution in buildings. It is the ratio between interior illuminance (from both direct and reflected light) and the illuminance of an outdoor unshaded plane under uniformly overcast sky. The daylight factor counts on the effect of glazing, air polluting, indoor and/or outdoor shading, etc. The daylight factor (in percent values) is determined from the relation:

$$D = \frac{E}{E_V}.100\,[\%] \tag{1}$$

where E is the illuminance of the reference interior plane [lx] and $E_V$ is reference illuminance in a point of the outdoor unshaded plane [lx] [2].

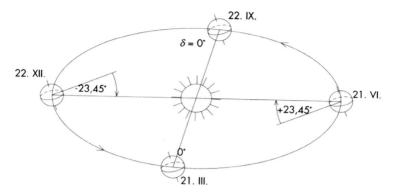

**Fig. 1** Solar declination during the year [2]

## 2.2 Declination of the Sun

Declination is the angle formed between sunrays and the terrestrial equatorial plane. The designation is δ. This parameter changes constantly during the year. For the purposes of this model, the declination is determined through mathematical relations.

$$\delta = 23,45° \sin\left[\frac{360°}{365.(J-81)}\right] \quad [°],$$ (2)

where J is the serial number of the specific day in a year [2] (Fig. 1).

## 2.3 The Length of the Solar Day

The solar length of a day is the traditional and most widespread time unit on the earth. The local solar time is defined as the mean of solar time.

Defined regions have their unified time. Central European Time (CET) coincides with the mean solar time at $\lambda_z = 15°$ east longitude and it is postulated as the zonal clock time for the longitude region $7.5° \leq \lambda_z \leq 22.5°$. True solar time can be determined from the Central European Time by using the relation:

$$GMT = UTC + \frac{\lambda_z - 15°}{15°} + ET \quad [\text{in hours}],$$ (3)

where $\lambda_Z$ is longitude (in degrees), GMT is the solar time (in hours), UTC is central European time (in hours), and ET is the time difference (in hours) [2].

## 2.4   Solar Elevation

Solar elevation $\gamma_S$ is the angle formed between the sunray and the horizontal plane. Solar elevation can be calculated by using the well-known formula:

$$\gamma_S = \arcsin[\sin\varphi.\sin\delta - \cos\varphi.\cos\gamma.\cos(15°GMT)] \ [°] \tag{4}$$

## 2.5   Diffuse Illuminance

Diffuse illuminance $D_V$, which is the illuminance of an unshaded outdoor plane under uniformly overcast sky that can be determined, based on the knowledge of the particular geographical and time coordinates, by using the following formula:

$$D_V = \left(\frac{D_{Vm}}{E_V}\right).E_{V0}.\varepsilon.\sin\gamma_S \ [\text{lx}], \tag{5}$$

where $D_{Vm}/E_V$ is the skylight transparency factor [2, 3].

## 3   Verification of the Prediction Model

Apart from mathematical calculations of the daylight and artificial light, verification of the model required long-term measurements of the electrical parameters in selected interior areas equipped with lighting systems dimmable to a constant illumination level. Taking into account the outdoor illuminance data for the unshaded plane at the building's geographical coordinates, the calculated daylight factor values are converted to the expected daylight contributions (in lx) of the daily illumination in the rooms that are being examined [4].

Tested room is located on the 2nd (above-ground) floor of the detached house at the VŠB—Technical University of Ostrava. The shading of surrounding buildings and trees were included in the calculation. The area of the room is 3975 mm × 3372 mm, and its height is 2100 mm. The shorter wall accommodates two windows which are 120 mm apart: the first window is 900 mm × 1000 mm area, at a height of 810 mm above the floor and 500 mm from the wall, and the second window is also 900 mm × 1000 mm area, at a height of 810 mm above the floor and 1520 mm from the wall.

The ground plant of the room (also indicating the locations of the windows) is obtained as output of the software for daylight calculations (in accordance with the applicable Czech standard) is shown in Fig. 2.

The artificial lighting system for our examined room must be designed to enable the power savings to be calculated. To meet the requirements in each computational

**Fig. 2** Daylight factor
calculations of tested room

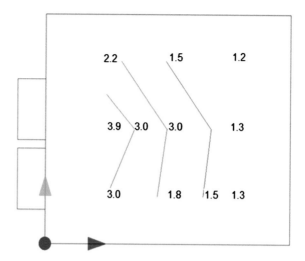

**Fig. 3** Artificial lighting
system calculation of the
tested room—illuminance

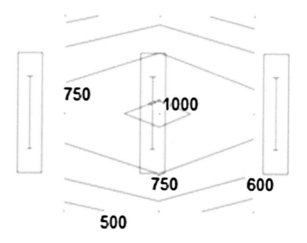

point, calculations must be performed according to the norm EN 12464-1. The lighting system calculations for the areas under study provide the maximum power input of the lighting system from which the power consumption by the non-dimmable lighting system can be obtained for the postulated use of the area. Room is equipped with three ceiling-mounted ZC T5 228/12LOS ZK, 2 × 28 W/840 luminaires [5, 6].(Figs. 3 and 4).

**Fig. 4** Interior of the tested room

# 4 Prediction Model Application

Once the time dependences of the illuminance from daylight (in the uniformly overcast sky conditions) are known for the space treated, a cycle of dimming calculations for the parts of the lighting systems (i.e. individual luminaires or light lines) can be launched based on the requirement that meet the minimum regulatory illuminance levels in the area. So the illumination levels (or permissible degrees of dimming) of the lighting system are obtained.

## 4.1 Electrical Energy Consumption of Lighting System

Electrical energy consumption of the lighting system depends on the lighting intensity of the room, whose values depend on the outdoor diffuse illuminance that determines the level of power demand of the lighting system which we need to achieve the required level of illumination.

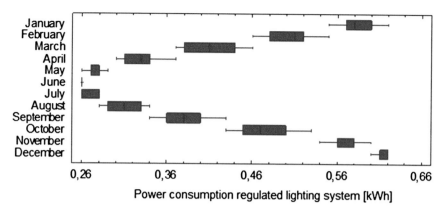

**Fig. 5** Power consumption of regulated lighting system in tested room

**Fig. 6** Consumption of regulated lighting system in the month of September

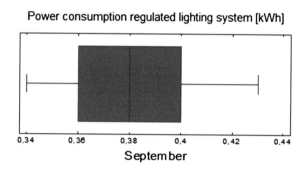

Figure 5: You can see that it is necessary to increase or reduce the energy consumptions according to individual months of a year in the tested room. The box diagram shows minimum, average and maximum levels of energy consumption throughout the year. These levels of consumption are influenced by intensity of cloudiness and daylight. Every month, the minimum rates of electrical consumption were measured at the time of smallest cloudiness and, in the contrary, the highest consumption was measured at the time of high level cloudiness [7].

There are the highest changes of consumption during winter and autumn months. The intensity of artificial lighting significantly fluctuates during the day because of changing levels of the daylight. That is why it is necessary to regulate the intensity of artificial lighting during the day in the winter time, and in the contrary it is possible to keep an approximate stable value of the artificial lighting in the summer due to constant daylight conditions. In the higher daylight intensity daytime it is possible to regulate consumption of artificial lighting system to minimum. The power demanded by artificial lighting system increases in autumn and winter months in comparison to the power demanded in spring and summer months, that means the power consumption is increasing as well.

Figure 6: We can find a lot of data in this box diagram. The minimum value of this diagram represents the lowest day consumption in the month, which is

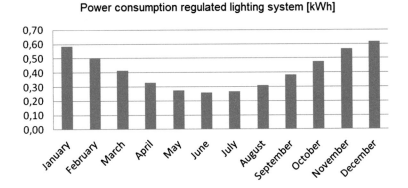

**Fig. 7** Power consumption of the regulated lighting system

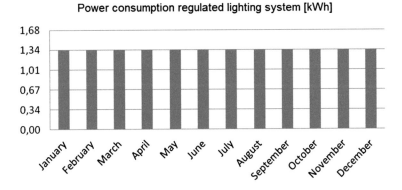

**Fig. 8** Power consumption of the unregulated lighting system

0.34 kWh. The maximum value represents the highest day consumption, which is 0.43 kWh. The average day consumption of this month is 0.38 kWh. The minimum values represent the electrical consumption on a clear day, while the maximum values show the consumption of electricity on a cloudy day. On the Fig. 5 it is evident that the consumption of regulated lighting systems changes according to year seasons, changing daylight time, sun height, etc. adequately [7].

Figures 7 and 8: Levels of electrical consumption of regulated and unregulated lighting systems are compared. In case of unregulated lighting system there is a constant level of electrical consumption of 1.34 kWh. In case of regulated lighting system there is the highest consumption in winter time, it is 0.62 kWh, and the lowest level of consumption in summer months, it is 0.26 kWh. In comparison to the unregulated lighting system there is very significant energy savings.

Figure 9: The importance of regulation of lighting systems in tested room is evident in this diagram. The regulation is able to save up to 69 % electric energy in our case.

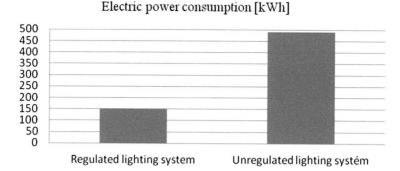

**Fig. 9** Electric power consumption

## 5 Conclusion

This article describes the application of the predictive model of the lighting system using natural daylight. From the statistical point of view it is determined that the regulation of lighting systems can save up to 69 % of electric energy. Total consumption of unregulated lighting system is 489 kWh/year, and on the other hand, the consumption of regulated lighting system is 151 kWh/year only.

The total yearly consumption of the artificial lighting system in tested room corresponds with the predictive model.

The measured values are influenced by the intensity level of daylight and cloud cover. Every day of a year the minimum consumption was measured on a clear day, while the highest consumption was at time of high level of cloudiness.

The intensity of artificial light significantly fluctuates during the day because of changing level of daylight during winter and autumn months. That is why it is necessary to regulate the intensity of artificial lighting during the day in winter time. On the contrary it is possible to keep approximate stable value of the artificial lighting in summer months due to constant daylight conditions. Therefore we are able to regulate lighting systems and achieve the desired illuminance values, reducing the power and consumption of artificial lighting and significantly increase energy savings.

**Acknowledgment** This article was prepared with the support of the project "Transmission of control signals in lighting systems" SP2015/182, by institution of VSB-TU Ostrava.

## References

1. Novak, T., Vanus, J., Sumpich, J., Koziorek, J., Sokansky, K., Hrbac, R.: Possibility to achieve the energy savings by the light control in smart home. In: Proceedings of the 7th International Scientific Symposium on Electrical Power Engineering, pp. 260–263 (2013)

2. Rybar, P., et al.: Denní osvětlení a oslunění budov, ERA 2001. ISBN 80-86517-33-0
3. Darula, S., et al.: Osvětlování světlovody, Grada Publishing 2009. ISBN 978-80-247-2459-1
4. Sumpich, J., Sokansky, K., Novak, T., Carbol, Z.: Potenciál úspor elektrické energie spotřebované soustavami sdruženého osvětlení s využitím denního světla, Electric power engineering, Brno, s. 1165–1168 (2012). ISBN 978-80-214-4514-7
5. Skoda, J., Baxant, P.: Control of Lighting Systems Using Compact Systems, Brno Univ Technology, pp. 797–798 (2010). ISBN 978-80-214-4094-4
6. ČSN EN 12464-1 Světlo a osvětlení—Osvětlení pracovních prostorů—Část 1: Vnitřní pracovní prostory
7. Software STATGRAPHICS 5.1. http://www.statgraphics.com/statgraphics_plus.htm

# User Identification by Biometric Methods

**Pavlina Nudzikova and Zdenek Slanina**

**Abstract** The concept of user identification by biometric methods is described. The paper summarizes techniques that are routinely used in biometric identification, gives the reader basic idea of the background of biometrics and explains the meaning of basic terms used in this area. The objective is to present an overview of existing methods for the recognition of people, which will be used in future research, particularly with focus on the recognition of individuals in the automotive industry.

**Keywords** Biometrics · Identification · Recognition · Fingerprint identification · Gait identification

## 1 Introduction

Biometrics is an automated method serving to verify the identity of a person based on recognition of the individual's unique biological characteristics. It is based on the belief that the biological characteristics are unique in human population and constant in time.

Anatomic or physiological properties are used for identification purposes. To be usable in biometrics, the characteristics must be unique, invariable, practically measurable and amenable to subsequent processing for evaluation by comparison with the same characteristics of other individuals. Biometrics is used to identify a person (identification) or to verify a person's identity (authentication) (Fig. 1).

Although biometrics has been in the focus recently particularly in relation to computer security, recognition by using biological characteristics has been used for

P. Nudzikova · Z. Slanina (✉)
Department of Cybernetics and Computer Science, Faculty of Electric Engineering and Computer Science, VSB—Technical University of Ostrava, Ostrava, Czech Republic
e-mail: zdenek.slanina@vsb.cz

P. Nudzikova
e-mail: pavlina.nudzikova@vsb.cz

© Springer International Publishing Switzerland 2016
V. Styskala et al. (eds.), *Intelligent Systems for Computer Modelling,*
Advances in Intelligent Systems and Computing 423,
DOI 10.1007/978-3-319-27644-1_17

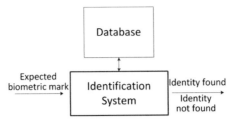

**Fig. 1** Principle of identification

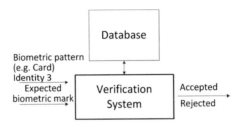

**Fig. 2** Principle of authentication

centuries or actually, millennia. Palm prints as sort of author's signature have been found in archaeological caves. Such prints are as much as 30,000 years old. Humans are generally able to tell one individual apart from another based on the individual's physiological features, typically the face and voice. However, it was not until the late 1960s that biometrics of humans started to be automated owing to the development of computer techniques (Fig. 2).

Biometrics of humans is superior to other means such as passwords or chip cards in that the characteristics cannot be lost or forgotten.

Also, biometric means of authentication are simple to use: it suffices to put a finger or palm of one's hand on a device or to look into a camera. As another asset, it is very difficult—actually nearly impossible—to falsify the data that are required for identification. Also, one individual's biometric characteristics cannot be transferred to any other individual.

The use of biometrics has also its drawbacks, though. While you can always tell if a digital password is valid or not, this is never that unambiguous in biometrics. A certain fraction of false identifications, although very very small, always exists. Biometrics is not, and never will be, absolutely reliable. This is so because one will never be able to provide a sample which is absolutely identical with that stored in the template.

Biometric systems are used in a number of industrial areas, among them automotive systems. Biometrics is used in active security systems, for driver and passenger authorization, in comfort systems, systems monitoring the driver for any signs of fatigue and systems to detect situations where the driver fails to pay full attention to driving. For use of biometrics in vehicles see Ref. [1].

## 2 Basic Methods of the Biometrics of Humans

Biometric characteristics can be categorized into anatomic-physiological characteristics and behavioral characteristics [2, 3]. Routinely used anatomic-physiological characteristics include fingerprints and prints of the palm of the hand, hand shape geometry and hand vein scans. Among behavioral biometric characteristics is, for example, keyboard/keypad pressing dynamics.

## 3 Biometric Feature Scanning

A biometric system is basically a recognition system which acquires biometric data from the individual logging into it. The system extracts a set of characteristic features from the information obtained and compares it to data stored in the database. Subsequently the system responds in some way based on the result of the comparison.

The scanning module scans a real analog input, which may be visual, thermal, voice, etc. This analog input is transformed into the digital form, i.e. binary data. It is at this stage that errors arise most frequently. Errors may be due to the effects of the environment (e.g. a dusty or scratched sensor) or to some inaccuracy on the scanning equipment side.

The characteristic feature is extracted from the scan in the extraction module. Not all of the information scanned is used to identify an individual: in fact, some important parts only are employed.

The feature is transmitted to the comparison module, where it is compared to the template, stored in the database of the current reference patterns. The result of the output is unique and should be acquired reasonably rapidly. This is related to the interconnection between the database and the scanner (and additional computers where applicable).

The final decision as to whether the data scanned match (are identical with) those stored in the database is made by the decision module (Fig. 3).

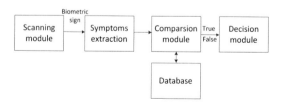

**Fig. 3** Biometric system modules

# 4 Biometrics of the Eye

Identification of humans through their retina and/or iris is among established bio-metric methods [4, 5], based on the fact that those parts of the human eye are unique in every individual. Identification methods based on those properties are highly reliable.

## 4.1 Recognition Through the Iris

This is a method that enables reliable identification of a person based on the unique reaction of the individual's iris to incident light. The iris has a random appearance, with a physical structure which is extremely rich in data patterns that differ from one person to another, including monozygotic twins. The process of user identifi-cation through iris recognition is shown in Fig. 4. The steps and methods are described in Refs. [6, 7]. Methods for iris location in the face are described in Ref. [8]. An optimized Daugman's algorithm for iris location is described in the paper by Hebaishy [9]. A comparison between Daugman's algorithm and the Hough trans-form is presented in Ref. [10]. A newer iris recognition method through mutual information estimation is introduced in Ref. [11].

## 4.2 Recognition Through the Retina

Recognition through the retina is a biometric method that compares individuals based on scanning and comparison of their retinal patterns. A special optical camera is required to obtain the image of the retinal veins. The retina is scanned by a low-intensity infrared beam which follows the circular path of the retina passing through the pupil [12]. The retina is more or less transparent to light at the wavelengths used. The image of the retina is actually formed only by the blood vessel network behind the retina, and it is that image that is used for the recognition of humans. The scanner beam runs around a certain area of the retina and captures a

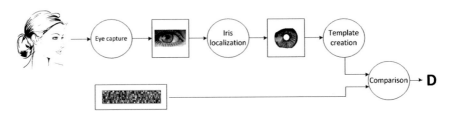

**Fig. 4** User identification by iris recognition [6]

circular picture of the retina where the vein structures are darker than the sur-rounding tissue. The methods to acquire and process the image are described in Refs. [4, 5].

# 5   Identification Through the Fingerprint

This method is based on scanning the person's fingerprint and its comparison with the database. The fingerprint examination-based method has the longest history and has found application in civil as well as forensic applications. Every person has their own unique fingerprints. The fingerprint pattern is even different in monozygotic twins (Fig. 5).

The pattern consists of papillary lines [6, 13] which are present on the finger pads. Algorithms that are currently applied to the identification of persons based on their fingerprints are described in Ref. [14]. Fingerprints are taken by means of various sensor devices, contact or contactless, whose function may be based on different physical principles [4, 5, 13]. Their ever more widespread use is facilitated by miniaturisation.

## 5.1   Verifying that the Source Is a Live Person

It is very important to make sure that a fingerprint arises from a live individual. In fact, people leave their fingerprints nearly everywhere, and so it is not difficult for one person to copy fingerprints left by another person. Therefore, when examining a fingerprint it is important to be sure that the fingerprint scanned is shown by the individual to whom it belongs and not by another person with the intent to deceive the system. Many methods to check that a fingerprint belongs to a live person exist. One of them is based on the detection of sweat (activity of the sweat pores). Another method uses the spectroscopic properties of human skin, additional methods are based on physical properties such as temperature, hot/cold stimuli, changes during pressing, pulse, etc.

**Fig. 5**  Fingerprint [5]

## 5.2   Use of the Fingerprint in a Car

Fingerprint based methods are very reliable and widespread particularly owing to the constant nature of the fingerprint structure and, importantly, low investment costs compared to other approaches. Fingerprint scanning equipment is routinely installed now into mass production cars. Typically, such equipment is used to identify the user before he or she starts the engine. The fingerprint can also be used for appropriate setting of the car interior parameters such as driver seat position, steering wheel position and air conditioning. The driver's favourite radio station may also be switched on.

Many car types are currently equipped with fingerprint scanning equipment which is used for identification combined with additional functions (typically security).

## 6   Biometrics of the Face and Face Parts

Identification of individuals through face recognition is the oldest method. People have been intuitively recognising their relatives, friends and other contacts for millennia.

Recognition of other persons based on their facial features has been the most natural and established method during the history of mankind. We perform this "operation" routinely during the day with very high reliability. It is natural, then, that efforts to achieve automatic recognition based on human face characteristics are among prominent ones in biometrics.

Software tools capable of recognising human face in the same manner as human brain does have been developed [15] (Fig. 6).

The software is based on various algorithms and methods for face identification [4, 5, 6]. It is a major advantage of this approach that the individual and the scanner need not be in direct contact. Still, many areas exist where the method can still be improved, for example, the scanner might also record signs of emotions. Continuous face scanning may be beneficial from the security aspect because the individual's authentication process would be repeated again and again rather than made within the signing-in procedure only (Fig. 7).

**Fig. 6** Biometrics of the face [5]

 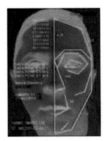

**Fig. 7** The different approaches to machine processing of the human face [5]

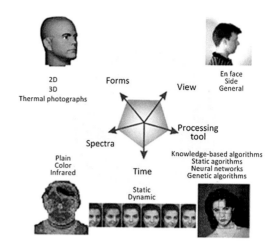

# 7  Gait Biometrics

Bipedal locomotion, or walking dynamics, is a next parameter usable to identify individuals. Locomotion is defined as the act or power of moving from place to place. Specifically in humans, locomotion is the motion of an individual in the gravitation field, driven by the individual's own power using the limbs or other anatomic parts of the body. Recognition of humans based on their gait is one of the newly emerging areas of biometric applications. For the principle of the method see Refs. [5, 16, 17]. Horák and Richter [18] achieved image segmentation through motion detection and by using the mean displacement method for the clustering of points. The segmentation methods are also described in Ref. [19] (Fig. 8).

**Fig. 8** Motion of human body's centre of gravity during walk. The trajectory is shown on the *right*. The *arrows* indicate maxima and minima [5]

**Table 1** Comparison of biometric methods

| Biometrics | Precision | Cost | Variability in time | User friendliness | Total |
|---|---|---|---|---|---|
| Fingerprint | ●●● | ●○○ | ●○○ | ●●○ | ●●● |
| Hand geometry | ●●○ | ●●○ | ●●○ | ●●○ | ●●○ |
| Face recognition | ●●○ | ●●○ | ●●○ | ●○○ | ●●○ |
| Iris recognition | ●●● | ●●● | ●○○ | ●○○ | ●●● |
| Retina recognition | ●●● | ●●● | ●●○ | ●●● | ●●○ |
| Nail bed | ●●● | ●●○ | ●●○ | ●●○ | ●●○ |
| DNA | ●●● | ●●● | ●○○ | ●●● | ●●○ |
| Voice recognition | ●○○ | ●○○ | ●●● | ●○○ | ●●○ |
| Signature dynamics | ●○○ | ●○○ | ●●○ | ●○○ | ●●○ |
| Key pressing dynamics | ●●○ | ●○○ | ●○○ | ●○○ | ●●○ |

Level—*Low* ●○○, *Medium* ●●○, *High* ●●●

## 8 Other Biometric Methods

One of the remaining biometric methods is vein biometrics [4]. The shape of the veins is measured when scanning the hand (either the back or the palm of the hand).

Additional methods include the use of the DNA, voice features, signature dynamics, keyboard typing dynamics, computer mouse moving dynamics, shape of the auricle, and others [4, 5, 6].

## 9 Comparison of the Biometric Methods

The basic biometric methods and their properties are compared in Table 1 [6]. The table demonstrates that the most reliable techniques include the fingerprint method, which is also most beneficial from the financial aspect and is user friendly.

## 10 Conclusion

Identity verification based on an individual's biometric properties is a rapidly developing field of science and technology [20], with a constantly increasing potential, number of methods and their combinations. The use of biometrics in the automotive industry is a current topic of interest. The methods are only at the onset of their development, and it will certainly be very interesting to engage oneself in this field of application for the car and truck industry. Various application areas are open to use, from inexpensive and simple ones for basic verification purposes to very sophisticated ones, and it is only up to you to choose the option that suits you best.

Combination of two different biometric methods is often advantageous: the rate of success of the recognition process is thus increased and it is easier to continue the authentication procedure if one of the scanning devices fails. Security is also enhanced: the double check makes it much more difficult to deceive the system.

Based on the knowledge acquired I decided to use the gait recognition method, which will pre-identify the user at the electromobile charging stations, followed by a more reliable method, such as the fingerprint scanning method where the scan match test uses a smaller selection of patterns stored in the database. Hence, a combination of two different methods will be used for the identification. The benefits of this approach include rapid user identification as well as enhanced identification precision. Whether the specific biometrical signs selected are suitable for identity check will depend on different criteria, especially the following: (i) Every individual must have a specific feature; (ii) it must be possible to discriminate between 2 individuals with adequate reliability based on that feature; (iii) the feature should be invariable or only little variable in time; and (iv) it must be possible to capture the feature by means of a detecting (scanning) device. Other technical aspects, which are also important, exist as well. They include the following: (i) the biometric feature capture and evaluation process should be adequately rapid; (ii) the system should exhibit adequate recognition precision and robustness in different conditions of use; (iii) the system should protect personal data against intrusion; (iv) the system should be able to disclose any attempt to use false biometric features. Future research will be aimed at examining the aspects with focus on whether it is just the combination selected that is most suitable for the identification of individuals especially in the automotive industry.

**Acknowledgements** This work is supported by project SP2014/188, named "Control of technological systems with OAZE providing an independent sustainable development of complex systems" of Student Grant Agency (VSB—Technical University of Ostrava).

# References

1. Plíhal, J.: Využití biometrie ve vozidle (Use of biometrics in vehicles). In: Proceedings of the International Professional Conference Improving the Safety of Vehicle Operation in Armed Forces (Zvýšení bezpečnosti provozu vozidel ozbrojených sil), pp. 93–98 (2011). ISBN 978-80-904625-2-6
2. Jain, A.K., Flynn, P., Ross, A.: Handbook of Biometrics, pp. 1–2. Springer, Berlin (2008). ISBN 978-0-387-71040-2
3. Denning, D.E.: Cryptography and Data Security. Addison Wesley, Boston (1982). ISBN 0-201-10150-5
4. Drahanský, M., Orság, F., et al.: Biometrie, Brno 2011, 1st edn. ISBN 978-80-254-8979-6
5. Rak, R., Matyáš, V., Říha, Z., et al.: Biometrie a identita člověka. Grada Publishing, Prague (2008). ISBN 978-80-247-2365-5
6. Núdziková, P., Slanina, Z., Vala, D., Drábek, P.: Electromobility I—User Identification (2014), ISBN 978-80-248-3531-0
7. Daugman, J.: How iris recognition works. IEEE Trans. CSVT **14**(1), 21–30 (2004). doi:10.1109/TCSVT.2003.818350

8. Jaskovský, P.: Person identification using iris recognition. Master's thesis, Czech Technical University, Prague (2013)
9. Hebaishy, M.A.: POSTER: Optimized Daugman's Algorithm for Iris. National Authority for Remote Sensing and Space Science, Egypt, Cairo (2004)
10. Mehta, H.R., Mehta, R.G.: Compare the techniques for iris localization. J. Inf. Knowl. Res. Electron. Commun. Eng. **2**(02), 666-670 (2012–2013). ISSN 0975-6779
11. Dobeš, M.: Image processing: new methods of localization and recognition human's iris. Dissertation, VSB-Technical University of Ostrava (2011)
12. Lichanska, A., Retina and iris scans. In: Encyclopedia of Espionage, Intelligence, and Security, p. 20. The Gale Group, Inc.
13. Kazik, M.: Fingerprint processing. Master's thesis, VUT Brno (2011)
14. Kovac, P.: Project of biometrical identification system for small organization. Master's thesis, UTB Zlin (2009)
15. Brunelli, R., Poggio, T.: Face recognition: features versus templates. IEEE Trans. PAMI **10** (15), 1042–1052 (1993)
16. Nixon, M.S., Carter, J.N.: Automatic recognition by gait. Proc. IEEE **94**(11), 2013–2024 (2006). doi:10.1109/JPROC.2006.886018
17. Choudhury, S.D., Tjahjadi, T.: Gait recognition based on shape and motion analysis of silhouette contours. Comput. Vis. Image Underst. **117**(12), 1770–1785 (2013). doi:10.1016/j.cviu.2013.08.003
18. Horák, K., Richter, M.: Segmentace obrazu pro identifikaci osob pomocí bipedální lokomoce. (Image segmentation to identify persons using bipedal locomotion). In: Proceedings of the International Conference Technical Computing, Prague 2009, p 36. HUMUSOFT, Prague (2009). ISBN 978-80-7080-733-0
19. Kryzanek, J.: Person identification by means of bipedal locomotion. Master's thesis, VUT Brno (2010)
20. Otahalova, T., Slanina, Z., Vala, D.: Embedded sensors system for real time biomedical data acquisition and analysis. In: 11th IFAC/IEEE International Conference on Programmable Devices and Embedded Systems, PDES 2012. WOS:000344655200050. ISBN 978-390282314-4, ISSN 14746670

# Energy Modelling of the Building

Jaroslav Kultan

**Abstract** The paper is focused on the modelling of energy consumption of buildings depending on outside weather and depending on the insulation of buildings. Based on the values of the outside temperature changes, schedule room-temperature set point and the measurement of the energy supply to the building it is possible to calculate the energy performance of buildings. In the calculations, it is necessary to consider not only the energy entering a building but also all the internal resources of the individual consumers as well as energy radiating from individuals living in the building. It is also necessary to take into account the intensity of solar radiation and also wind speed and direction. On the basis of precise measurements, it is possible to create a model of the building in terms of energy consumption, which can be compared with the theoretical losses imposed by the technical parameters of the building. This model can be used to actively control the operation of power equipment in a building according to the outside weather and the needs of the population. Based on the parameters it is possible to design a system of basic and additional heating based on renewable resources or recuperative technologies.

**Keywords** Energy consumption · The energy model of the building · Loss of energy · Renewable energy · Additional resources

## 1 Introduction

One of the main problems of building maintenance is their efficient use, in compliance with the required temperature regime. They must comply with all conditions sufficient energy not only in winter but also dissipate heat from the building during the summer. Energy supply, not only in industry but also in public services is one of

J. Kultan (✉)
Faculty of Economic Informatics, Department of Applied Informatics,
University of Economics in Bratislava, Bratislava, Slovak Republic
e-mail: jkultan@gmail.com

© Springer International Publishing Switzerland 2016
V. Styskala et al. (eds.), *Intelligent Systems for Computer Modelling*,
Advances in Intelligent Systems and Computing 423,
DOI 10.1007/978-3-319-27644-1_18

the main tasks. The cost of energy required for full operation of buildings is one of the existing components of current expenditure. The main objective of this paper is an energy model and existing buildings on the basis of lessons learned highlight the potential for renewable energy [1].

From the measured values of energy consumption will be made energy model of the building. This model will be used to analyze the possibility of the use of power equipment to improve the energy balance of the building.

## 2  Creating Energy Model

The energy model of the building can be obtained in several ways. One is the creation of a model by measuring the power consumption and obtaining additional energy from man-made or the external environment. The other way is to create a model based on the parameters of buildings, various building materials and the temperature gradient between the room temperature and outside temperature.

To create a model based on consumption measurements

In this case, assume that a particular building is a black box on the values of input and output variables, it is possible to create a model of the object.

By measuring the energy inputs assume an energy model type

$$\mathbf{y} = \mathbf{A}^* \mathbf{x} \qquad (1)$$

where $\mathbf{y}$—is the measured energy consumed in buildings, $\mathbf{A}$—energy model of the building, which characterizes the dependence of energy from the outside temperature, $\mathbf{x}$—is the measured outside temperature.

If input quantities are several types of energy and measures the number of external parameters/temperature, wind speed and direction, solar radiation intensity/given model has the form:

$$\mathbf{Y} = \mathbf{A}^* \mathbf{X} \qquad (2)$$

where the vector $\mathbf{Y} = (y_1, y_2, \dots y_n)$ is the vector of input energy—calculated on the single unit e.g., kWh; $\mathbf{X} = (x_1, x_2, \dots x_m)$ is a vector of measured variables of weather, for example. outside temperature, wind speed, wind direction, solar radiation intensity, humidity, pressure and so on, $\mathbf{A}$—energy model of the building

This model does not always accurately reflect the dynamic changes in energy dependence from changes in weather parameters. Large deviations of model outputs and real systems may arise when rapid changes in weather parameters. Certain variables, for example outside temperature, not too rapid changes in its value. Therefore there is no need to bring in the model and previous values. A greater impact on power consumption changes can have a change of wind speed or change in intensity of solar radiation. The house has a relatively high inertia and therefore consumption energy depending on the wind speed and direction may not be sufficiently precise. Accurate results showed a model that takes into account current and passing the values of the wind speed. A similar impact on consumption may also change in the intensity of solar radiation. A rapid increase in intensity in cloudy weather may not result in a rapid change in supply of energy.

It is therefore preferred to create dynamic energy model that takes into account the last power state/output/and the last state of some selected weather parameters/input.

It is therefore preferred to create dynamic energy model that takes into account the past of consumption and the last state of some selected zones.

$$y(t+1) = A^*(y(t), x(t), x(t-1), x(t-2), \ldots x(t-i), \ldots), \tag{3}$$

Different types of input variables can have different temporal history.

We assume that the more of the species is possibly faster change, the more historical values should be placed in a model system.

If the house uses several types of energy/electricity, gas, hot water/from external sources (hot water—domestic hot water from a central system, thermal energy from the central distribution), we can create the following model type:

$$\begin{pmatrix} y_{1(t+1)} \\ \vdots \\ y_{n(t+1)} \end{pmatrix} = \begin{pmatrix} a_{1,1} & \cdots & a_{1,m} \\ \vdots & \ddots & \vdots \\ a_{n,m} & \cdots & a_{n,m} \end{pmatrix} * \begin{pmatrix} y_{1(t)} \\ \vdots \\ y_{1(t-m_1)} \\ \vdots \\ y_{n(t)} \\ \vdots \\ y_{n(t-m_n)} \\ \vdots \\ x_{1(t)} \\ x_{1(t-1)} \\ \vdots \\ x_{1(t-m_{n+1})} \\ \vdots \\ x_{k(t-m_{n+k})} \end{pmatrix} \tag{4}$$

where

n                    is the number of types of energy used;

k                    is the number of external parameters;

$m = \sum_{i=1}^{n+k} m_i$    the number of entries of A.

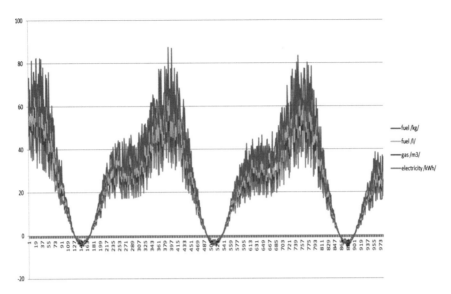

**Fig. 1** Individual consumption of energy in natural units three years [2]

Model (4) represents one possible, simple energy building model that allows us to calculate the expected performance of individual funds depending on the weather conditions and the state of the building.

Making energy model based on measured values of the energy (Fig. 1) over a period of time, preferably at least a year. However, the formation of the model may not be dependent on the number of measurement days, if we use an adaptive model, or at least the calculation of the adaptive algorithm of the model and the continuous modification.

In that house on the recovery of the heat is used for heating oil. During the transition period and in winter, heated by burning wood in the fireplace. In addition, the connection of gas boiler. The gas is also used to prepare the food in a. Electricity is mainly used for lighting, appliances at work, preparing the water when cooking or reheating water in the reservoirs.

The energy consumption of each resource is measured in different quantities, and therefore they need to calculate the single value—kWh (Fig. 2). For creating the model is suitable individual variables have the same dimensions.

Simultaneously measured weather parameters, which include e.g. wind speed, intensity of sunlight, humidity, pressure and so on. The graph (Fig. 3) is an example of the measured data to be used in creating the model [3]. This is the temperature, wind speed and direction and cloud.

For proper model creation it is necessary to record the number of residents in the apartment/house. Every person is a generator of thermal energy and have an impact on the overall energy balance.

For the purposes of model building it is therefore necessary to have the following information:

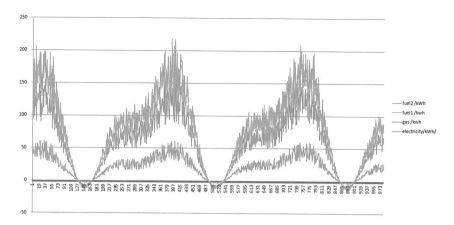

**Fig. 2** Individual energy consumption in kWh three years

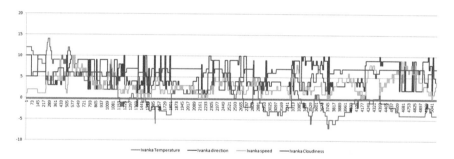

**Fig. 3** Measurements of weather parameters

$$E = (E_1, E_2, \ldots E_i, \ldots En, ) \tag{5}$$

$$P = ( P_1, P_2, \ldots P_i, \ldots Pm,) \tag{6}$$

$$Ob = \left( ob_1, ob_{2,}, \ldots, ob_{i,} \ldots \right) \tag{7}$$

$$T = (t, t - dt, t - 2^*dt, \ldots t - i^*dt \ldots) \tag{8}$$

where $E_i = (e_i(t), e_i(t{-}1), e_i(t{-}2) \ldots e_i(t{-}j)\ldots)$ the data on the consumption of energy types, $P_i = (p_i(t), p_i(t{-}1), p_i(t{-}2) \ldots p_i(t{-}j)\ldots)$ the data on the status of individual weather parameters, Ob—is a quantity that represents the number of persons in the house, T—vector of various time points, including the date when the variables measured.

Parameter dt is levied on the technical possibilities of inertia of the system and criteria of stability while driving.

The data on population affect the energy balance of the house and it is therefore necessary to create a vector which measures the number of residents in the house. This is a vector that needs to be formed as a time sequence.

Number of switching on the light, or other activities, and their power consumption is part of the vector Pi.

The calculation of the model and calculate the relationship:

$$Y = A^*X \tag{9}$$

$$\text{Where } Y = \begin{pmatrix} e_{1,1} & \cdots & e_{1,n} \\ \vdots & \ddots & \vdots \\ e_{k,1} & \cdots & e_{k,n} \end{pmatrix} \tag{10}$$

$$X = \begin{pmatrix} p_{1,1} & \cdots & p_{1,m} & ob_1 & t_1 \\ \vdots & \ddots & \vdots & \vdots & \vdots \\ p_{k,1} & \cdots & p_{k,m} & ob_k & t_k \end{pmatrix} \tag{11}$$

## 2.1   The approximation line segment

The energy consumption depending on the weather, can be considered a non-linear system. There are many nonlinear models of real systems, which could be used for computationally consumption depending on the weather.

In this post lists a relatively new model—a model of linear intervals.

To identify non-linear energy systems can also use the method with linear models [4], [5]. Such an approach is for example an interval linearization method. Nonlinear systems modeled as a set of linear models. Each linear model represents a nonlinear system in certain area—*area of linearization*.

This section shows the basic methods of linearization of an interval of two scalar variables—input quantity u(t) and the corresponding output variable y(t) obtained by measurement to identify non-linear system. At the beginning we choose the number of intervals of input variables **mu**, and number intervals of output variables **my**. Then select the limits of individual intervals $uh_r$ and $yh_s$, for $r = 0, 1, 2, \ldots mu$, $s = 0, 1, 2, \ldots my$. The boundaries defining the different sections of linearization $D_{r, s}$, which may also be referred to as $D^k$. Calculation serial number model using parameters r and s for systems with one input and output is given in (16). Non-linearity of the system is expressed in the form of linear Eq. (12) and the number of linearization segments is expressed in the Eq. (15).

Choice of thresholds $uh_r$ and $yh_s$ for $r = 0, 1, 2, \ldots$ it, $s = 0, 1, 2, \ldots$ we define intervals linearization $(uh_{r-1}-uh_r)$ and $(yh_{s-1}-yh_s)$ (from them name derived methods) that define the relevant *sections linearization* $D_{r, s}$, representing a $D^k$. Calculation serial number model using parameters races for systems with one input and output is given in (16) Model non-linear system is expressed in the form of a linear Eq. (12) and the total number of sections linearization is expressed in the Eq. (15)

Equivalent linear model system with an input and an output variable for the k-th interval linearization we express the difference equation

$$y(t) = q_0^k + \sum_{i=1}^{ny} q_i^k y(t - iT) + \sum_{j=1}^{nu} q_{ny+j}^k u(t - (j - d + 1)T) \qquad (12)$$

or a shorthand form

$$y(t) = z(t)q^k \quad k = 1, 2 \ldots, kk \qquad (13)$$

where

$$z(t) = [1, y(t - 1), \ldots, y(t - ny), u(t - d), \ldots, u(t - d - nu + 1)]$$

$q^k$ is a vector model system in the k-th interval linearization, while

$$qk = (q_0^k, q_1^k \ldots, q_{nu+ny}^k) \qquad (14)$$

Nu    number of starting variables entering the model
Ny    number of samples the output variables entering the model
T     time
T     sampling time
d     input variable displacement
k     index interval linearization

The total number of sections for linearization kk where nu = 1 and ny = 1 calculate

$$kk = mu.my \qquad (15)$$

index k—serial number linearization section express the relationship

$$k = (r - 1).my + s \qquad (16)$$

where there are more input quantities and more output quantities nu > 1 and ny > 1

$$kk = mu^{nu}.my^{ny} \qquad (17)$$

The value of index linearization is calculated as follows. If the output quantity y $(t - i)$, $i = 1, 2, \ldots, ny$ is the amplitude of the area $(yk_{s-1}, yk_s)$, $s = 1, 2, \ldots, my$, we then denote the character area with $I_i = s$. Similarly, if the value $u(t - j - d + 1)$, $j = 1, 2, \ldots, nu$ is from zone $(uk_{r-1}, uk_r)$, $r = 1, 2, \ldots, mu$, then the this area shall indicate the $I_{ny+j} = r$. The values of these intervals inserted into a vector $I = (I_1, I_2, \ldots I_{nu+ny})$. Index linearization interval calculated according to

$$k = I_1 + \sum_{l=2}^{nu+ny} (I_l - 1).\delta_l \qquad (18)$$

where

$\delta_1 = my^{(l-1)}$           for l=1, 2, ..., ny
$\delta_{l+ny} = my^{ny}.mu^{(l-1)}$   for l = 1, 2, ..., nu
l                   auxiliary index

From the measurement of the long-term operating u(t), y(t) we know the dependence of $f = f(u(t), y(t))$ generally does not pass all the possible areas of Dk, k = 1,2,..., kk. Therefore, it is usually necessary to construct a smaller number of linear models, such as the maximum number kk.

Construction of the linear model (12), respectively (13) for the k-th interval, the linearization is to determine the coefficient of linearization of the vector (14) from the measured values $y(t-i)$ and $u(t-j-d+1)$ for i = 1,..., ny, j = 1,... nu with the k-th interval linearization..

The model is obtained from the relationship $\mathbf{A}^k \mathbf{q}^k = \mathbf{b}^{k,}$ of which expresses the vector model $\mathbf{q}^k$ for k-section linearization method of least squares. It is necessary to create a matrix $\mathbf{A}^k$ from measured value $y(t-i)$ and $u(t-j-d+1)$ and vector $\mathbf{b}^k$ from measured value y(t).

For r-line of matrix $\mathbf{A}^k$ we can write

$$
\begin{aligned}
a^k_{r,i+1} &= y(t-i) && \text{for } i = 1, 2, .., ny \\
a^k_{r,j+1+ny} &= u(t-j-d+1) && \text{for } j = 1, 2, \ldots, nu \\
a^k_1 &= 1
\end{aligned}
$$

For element of vector $\mathbf{b}^k$

$$b^k_r = y(t-r)$$

where r = 1, 2, ... n; ny + nu + 1 < n

The results of such a simulation model is shown in Fig. 4.

The essential features of the method is the dependence of the accuracy of the identified model, the number of selected intervals, which divide the input and output signal, and the choice of the borders of the selected intervals. Their appropriate change we are trying to achieve the highest accuracy of models that quantify example. the sum of quadrates of deviations given patterns you-degrees values of the respective samples-originálu and the model output for a given interval of independent variable.

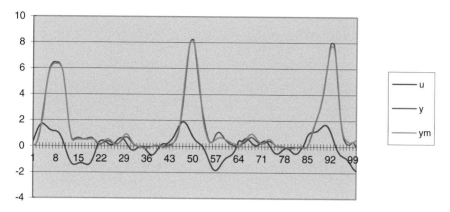

**Fig. 4** Modeling of energy consumption depending on external conditions

**Fig. 5** Distribution/
composition/wall in heat
transfer calculations

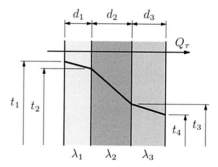

## 2.2   Energy model system based on physical properties

This model represents a system of equations and inequalities building on the basic
physical patterns on the transfer of heat between the flat panel composed of several
layers. In this case, it is necessary to know the thermal conductivity and thermal
resistance of the material, number of layers of which the house is composed. Then,
on the basis, according to such relations, it is possible to calculate the energy loss of
the house (Fig. 5).

These energy losses compared with the total loss of the house in accordance with
the commanding remove energy.

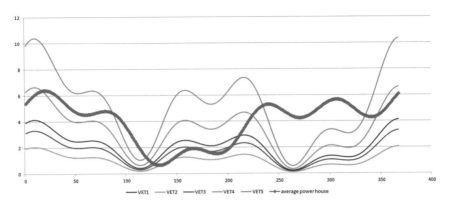

**Fig. 6** Comparison of household consumption and possibilities of energy production from wind power

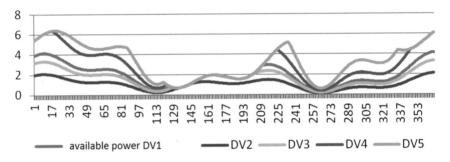

**Fig. 7** The projected amount of energy VET

## 3   Use of Models

The resulting mathematical model may be used in the design of additional energy sources for the building or the design of the switch from traditional sources for renewable energy. Based on weather data for the measured period can be calculated in available energy from various renewable sources. Based on the course during the design and the additional resource, it is possible to remodel cover the consumption of energy in the house. The information thus obtained are the basis for the design and implementation of such devices.

To reduce the energy dependence of the house from primary sources, it is possible to select multiple wind farms (Fig. 6) with different diameter rotor (VET1, VET2, …VET4), which differ not only the size but also the price of the connection method, operating costs, etc.

Based on the curves of energy production from wind and average consumption of the house during the whole year, you can choose a suitable VET so that each source to make the most. The following diagram (Fig. 7) is shown the available amount of energy that can deliver different sources that can absorb the house. These figures can be seen what part the energy can be covered with an energy source.

## 4   Conclusion

Creating energy model allows each building to carry out an analysis of the technical condition of buildings and the proposal to establish an effective additional energy source. On analyzing the technical condition to be compared created model energy system based on the input–output characteristics with model created by physical calculation of the house. In the event of a significant deviation can be assumed aging materials or damage insulation properties. For analysis of additional resources it is appropriate to calculate the estimated consumption of the house during a certain period and the possible energy yield and substantial financial savings for the introduction of selected types of renewable sources. Compare the cost of installing and operating the supply with energy savings can be calculated the overall economic return.

Creating energy model for individual houses is not the ultimate goal but a means to implement economically efficient energy management of buildings.

## References

1. Kultan, J.: Adaptívne riadenie spotreby. In: EE - časopis pre elektrotechniku, elektroenergetiku, informačné a komunikačné technológie [elektronický zdroj] : zborník ku konferencii Elektrotechnika, informatika a telekomunikácie 2013, 15–18 október 2013, Bratislava. - Bratislava : Spolok absolventov a priateľov FEI STU v Bratislave, 2013. - ISSN 1335-2547. - Roč. 19, mimoriadne číslo (október 2013), s. 162-164 CD-ROM
2. Kultan, J., Baitassov, T., Ishankulov, M., Rivkina, N.: In: Renewable Energy Sources 2013: Proceedings of the 4th International Scientific Conference OZE: Tatranské Matliare, Slovakia, pp. 247–253. 21–23 May 2013. Slovak University of Technology in Bratislava, Bratislava (2013). ISBN 978-80-89402-64-9
3. Kultan, J., Baitassov, T.: Energy management and renewable energy sources. In: Power Engineering 2012: Abstracts of 3rd International Conference Renewable Energy Sources 2012: Tatranské Matliare, Slovakia, pp. S. 241–242. 15–17 May 2012. Slovak University of Technology in Bratislava, Bratislava (2012). ISBN 978-80-89402-48-9
4. Kultan, J.: Nonlinear System Models Based on Interval Linearization a Medical Application. Medzinárodná konferencia IFFAC, Bratislava (2003)
5. Harsanyi, L., Kultan, J.: Method of selective forgetting for nonlinear system identification. J. Electr. Eng. **43**(7), 207–210, Bratislava (1992)

# Reconstruction of 4D CTA Brain Perfusion Images Using Transformation Methods

Iveta Bryjova, Jan Kubicek, Michal Dembowski, Michal Kodaj
and Marek Penhaker

**Abstract** The CT angiography (CTA) method is a mini-invasive diagnostic method for displaying blood vessels using computer tomography (CT) and the concurrent application of a contrast agent (CA). This article focuses on assessing brain perfusion in time based on a 4D reconstruction using one of the image transformation methods—morphing. The proposed methodology is very important for clinical practise. On the base this approach we are able to perform reconstruction of 4D CTA brain perfusion without using contrast substance. It is main difference against conventional procedures which are used during the examination. Patient is not exposed by contrast substance.

**Keywords** CT angiography (CTA) · Computer tomography (CT) · Contrast agent (CA) · Post-processing · Morphing · Brain perfusion · Image segmentation

I. Bryjova · J. Kubicek · M. Dembowski · M. Penhaker (✉)
The Department of Cybernetics and Biomedical Engineering, FEI, VSB-TU Ostrava,
17. listopadu 15, 708 33 Ostrava-Poruba, The Czech Republic
e-mail: marek.penhaker@vsb.cz

I. Bryjova
e-mail: iveta.bryjova@vsb.cz

J. Kubicek
e-mail: jan.kubicek@vsb.cz

M. Dembowski
e-mail: michal.dembowski@gmail.com

M. Kodaj
Podlesí Hospital, a. s., Konská 453, 739 61 Třinec, The Czech Republic
e-mail: michal.kodaj@gmail.cz

© Springer International Publishing Switzerland 2016
V. Styskala et al. (eds.), *Intelligent Systems for Computer Modelling*,
Advances in Intelligent Systems and Computing 423,
DOI 10.1007/978-3-319-27644-1_19

# 1 Introduction

CT angiography (CTA) is a mini-invasive diagnostic method based on computer tomography imaging technology that allows a user to display three-dimensional vascular structures onto one level called axial sections. The acquired data can then be processed and converted into a static three-dimensional shape (3D) display or kinetic display of space in time (4D). For a clinically sufficient display of vascular structures, it is important to achieve the optimal spatial resolution [1]. Perfusion brain examination is based on 2 series of images in the upper third of the basal ganglia and in the area above the basal ganglia. Perfusion values are calculated using specialized software to display perfusion maps that are used in clinical practice for diagnostic evaluation of brain perfusion [2–5] (Fig. 1).

# 2 Brain Perfusion

Perfusion examinations are intended for assessing blood flow and determining circulation time (the advance application of a contrast agent); however, they are not appropriate for assessing expansionary processes. Evaluating tissue perfusion is performed using colored perfusion maps, where the spots of blood circulation, or places with poor blood distribution, are clearly marked. Perfusion examinations use two axial sections at one position of a patient table for each phase and at a scan rate of 1 s. The result is two series of axial displays showing cerebral arteries saturated by the CA in the given section. Individual series are comprised of approximately 40 individual axial images (80 images in total for the entire perfusion examination). The first stage is aimed at the upper third of the basal ganglia and the other above the level of the basal ganglia [3, 6–10] (Fig. 2).

**Fig. 1** 4D brain perfusion imagining using TBV map [6]

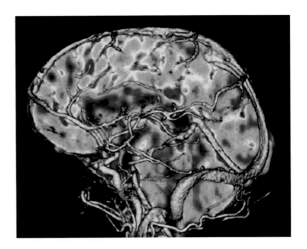

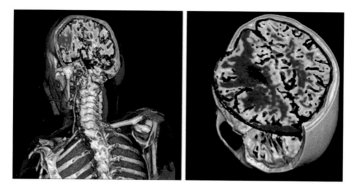

**Fig. 2** Imagining of 4D perfusion [6]

## 3 Design of Time Dependent Visualization of Brain Perfusion

To visualize dynamic brain perfusion over time, we used the morphing method. Morphing is a method that provides a smooth transition between the initial and final images, or a selected series of images. We applied this method to the series of axial images taken in the given section during the perfusion brain examination using the CTA method [11]. The aim was to create a continuous visualization of the intracranial filling of vessels with a CA over time and to indicate the location of possible insufficiency. The method allows for an accurate indication of the location of arterial occlusion and for drawing other consequences of insufficiency. Post-processing occurred with the help of the MATLAB® interactive programming environment, where we implemented morphing sequences with the Free Morphing® programme. This combination led to the creation of a functional user interface for further analysis of perfusion images. We have verified the results with available perfusion maps. The proposed algorithm is described by following steps:

- Defining of control points position.
- Defining of number control images.
- Defining of nth image generates interpolation site from images A and B of positions of intersections. This site is consequently applied on image B and we get image $B_n$. Same site is consequently applied on image A and we get image $A_n$. In the next step both images are added up. On the base of described procedure we get sequence of image animation.

Key element of algorithm is defining of sites which must be precisely suggested to avoid unwanted changes and artefacts. It is obvious, if we reach higher similarity of images we will get smoother transition among individual images. 3D morphing is processed in model space and usually is simpler for models with same number of defining elements [1, 12–15].

# 4   Brain Perfusion Image Processing (BPI)

Mapping algorithms for brain perfusion are implemented into software environment for analysis a processing of perfusion images. Main aim of this software solution is better indication and visualization of perfusion large intracranial artery. The proposed software is composed from three essential parts:

- Original perfusion—Area of original perfusion CT images or color CT maps.
- Processes perfusion—Area of visualization for post processing of selected perfusion images.
- Morphed perfusion—Area of visualization for created morphed sequence.

On the input of algorithm are color perfusion maps or axial CT images after applying contrast agent in given slide. Selected data are shown in graphical window Original perfusion. The proposed solution offers a few alternatives for post processing of perfusion images which are called in panel function. The output image data are shown in graphical window processed perfusion. Software also allows loading of sequence morphing perfusion examination. After taking this procedure is possible to run video sequence of perfusion within the time (Fig. 3).

The first option of image processing uses algorithm Color reverse. This procedure allows inversion of brightness scale for better recognition and detection of intracranial blood vessels which are fulfilled by contrast agent. Individual perfusion images are indicated by number 1 for bright pixels and 0 for dark pixels. Inversion of brightness scale is given by subtraction of individual pixels and thereby output image is formed (Fig. 4).

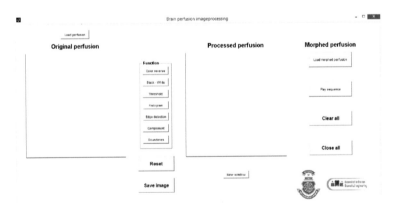

**Fig. 3**  Design of software environment for processing of perfusion images

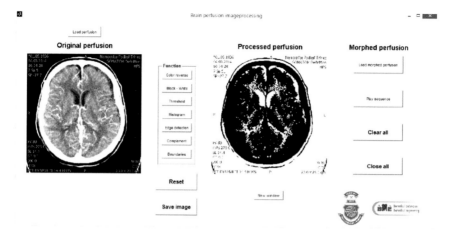

**Fig. 4** Comparison of brain perfusion (*left*) and algorithm color reverse (*right*)

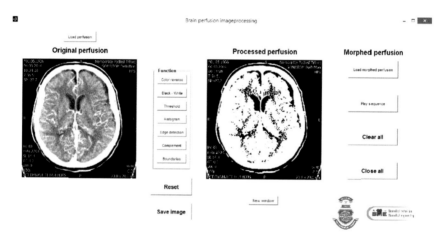

**Fig. 5** Comparison of brain perfusion (*left*) and image binarization (*right*)

Second alternative for processing of brain perfusion is binarization of image data. This imaging mode is especially appropriate for better imaging of intracerebral chambers and spots with lower brightness intensity of original perfusion image. This approach uses Otsu method and subsequent image transformation to binary appearance (Fig. 5).

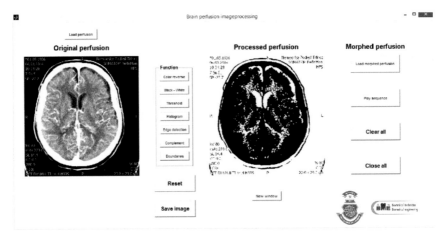

**Fig. 6** Comparison of brain perfusion (*left*) and thresholding of perfusion map (*right*)

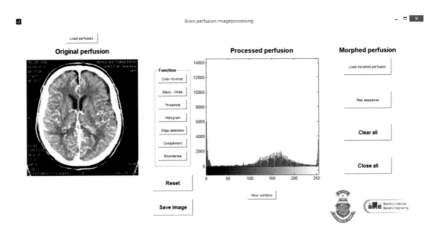

**Fig. 7** Comparison brain perfusion (*left*) and histogram analysis (*right*)

The last alternative for mapping perfusion images brings approach on the base of image thresholding. This algorithm unlike previous approach uses more thresholding levels. This algorithm offers appropriate alternative for selection certain brightness areas. Some areas can be suppressed and others can be highlighted (Fig. 6).

Software application also allows selection of individual thresholds directly from histogram analysis. In dependence on area which should be highlighted is possible to interactively set thresholding values. Histogram analysis is essential procedure in the area of processing perfusion images. It is necessary in order to showed intervals with the highest frequency were shown from middle tones up to brightest (Fig. 7).

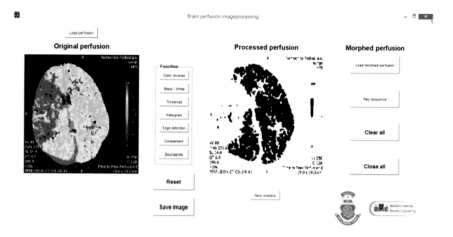

**Fig. 8** Comparison of original perfusion map TTP (*left*) and function complement (*right*)

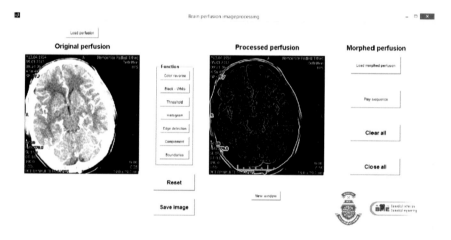

**Fig. 9** Comparison of brain perfusion (*left*) and structure segmentation (*right*)

Software equipment for mapping of brain perfusion allows two additional showing modes. During post processing original perfusion CT images intracranial blood vessels are shown. These areas are accompanied by the highest brightness. During post processing of color perfusion maps, only ideal perfusion brain tissues are shown. This fact leads to improving of detection pathological areas (Fig. 8).

For performing better diagnosis is important segmentation procedure which separates individual structures of perfusion image or color perfusion map. After taking post processing of perfusion CT brain tissues with suppressing intracranial blood vessels system is shown. In the case of color perfusion maps individuals perfusion areas in each showing mode is indicated. This segmentation function improves recognizability and indication of brain perfusion (Figs. 9 and 10).

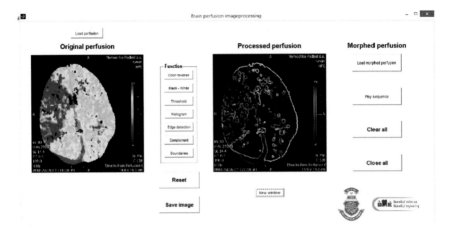

**Fig. 10** Comparison of perfusion map TTP (*left*) and structure segmentation (*right*)

# 5 Conclusion

Main aim of our research is design of software environment for analysis of 4D perfusion data. The user interface is composed from three essential parts. The first part offers possibility of visualization original perfusion image data. The second part contains segmentation outputs which serves for indications whole perfusion or identification of pathological areas on intracranial system. The third part of system serves for visualization of saved morphing sequence which are processed from same patients data which were used for image post processing. Development of software solution proceeds in cooperation with radiology experts and currently software is being tested in clinical practice.

**Acknowledgements** The work and the contributions were supported by the project SP2015/179 'Biomedicínské inženýrské systémy XI' and This work is partially supported by the Science and Research Fund 2014 of the Moravia-Silesian Region, Czech Republic and this paper has been elaborated in the framework of the project "Support research and development in the Moravian-Silesian Region 2014 DT 1—Research Teams" (RRC/07/2014). Financed from the budget of the Moravian-Silesian Region.

# References

1. Davis, B., Royalty, K., Kowarschik, M., Rohkohl, C., Oberstar, E., Aagaard-Kienitz, B., Niemann, D., Ozkan, O., Strother, C., MISTRETTA, C.: 4D digital subtraction angiography: implementation and demonstration of feasibility. Am. J. Neuroradiol. **34**(10), 1914–1921 (2013). doi:10.3174/ajnr.A3529
2. Penhaker, M., Matejka, V.: Image registration in neurology applications. In: International Conference on Networking and Information Technology (ICNIT), pp. 550–553 (2010)

3. Kasik, V., Penhaker, M., Novak, V., Pustkova, R., Kutalek, F.: Bio-inspired genetic algorithms on FPGA evolvable hardware. In: 4th Asian Conference on Intelligent Information and Database Systems, ACIIDS 2012. LNAI, vol. 7197, pp. 439–447, Kaohsiung (2012)

4. Baka, N., Metz, C.T., Schultz, C., Neefjes, L., Van Geuns, R.J., Lelieveldt, B.P.F., Niessen, W.J., Van Walsum, T., De Bruijne, M.: Statistical coronary motion models for 2D t/3D registration of X-ray coronary angiography and CTA. In: Medical Image Analysis, vol. 17(6), pp. 698–709 (2013)

5. Scardapane, A., Stabile Ianora, A., Sabbà, C., Moschetta, M., Suppressa, P., Castorani, L., Angelelli, G.: Dynamic 4D MR angiography versus multislice CT angiography in the evaluation of vascular hepatic involvement in hereditary haemorrhagic telangiectasia. Radiol. Med. **117**(1), 29–45 (2012). doi:10.1007/s11547-011-0688-3

6. Ferda, J.: CT angiografie. 1. vyd. Praha: Galén, 2004, xi, 408 s. ISBN 80-726-2281-1

7. Perfusion primer, http://neuroangio.org/neuroangio-topics/perfusion-primer/

8. Willems, P.W.A., Brouwer, P.A., Barfett, J.J., Terbrugge, K.G., KRINGS, T.: Detection and classification of cranial dural arteriovenous fistulas using 4D-CT angiography: initial experience. Am. J. Neuroradiol. doi:10.3174/ajnr.A2248

9. Kubicek, J., Penhaker, M., Pavelova, K., Selamat, A., Hudak, R., Majernik, J.: Segmentation of MRI data to extract the blood vessels based on fuzzy thresholding. In: New Trends in Intelligent Information and Database Systems, pp. 43–52. Springer International Publishing (2015)

10. Frölich, A.M.J., Wolff, S.L., Psychogios, M.N., Klotz, E., Schramm, R., Waser, K., Knauth, M., Schramm, P.: Time-resolved assessment of collateral flow using 4D CT angiography in large-vessel occlusion stroke. Eur. Radiol. **24**(2), 390–396 (2014). doi:10.1007/s00330-013-3024-6

11. Yamaguchi, S., Takeda, M., Mitsuhara, T., Kajihara, S., Mukada, K., Eguchi, K., Kajihara, Y., Takemoto, K., Sugiyama, K., Kurisu, K.: Application of 4D-CTA using 320-row area detector computed tomography on spinal arteriovenous fistulae: initial experience. Neurosurg. Rev. **36**(2), 289–296 (2013). doi:10.1007/s10143-012-0440-z

12. Pustkova, R., Kutalek, F., Penhaker, M., Novak, V.: Measurement and calculation of cerebrospinal fluid in proportion to the skull. In: 9th Roedunet International Conference (RoEduNet), pp. 95–99 (2010)

13. Mendrik, A., Vonken, E., Van Ginneken, B., Smit, E., Annet Waaijer, A., Bertolini, G., Viergever, M.A., Prokop, M.: Automatic segmentation of intracranial arteries and veins in four-dimensional cerebral CT perfusion scans. Med. Phys. **37**(6), 2956–2966 (2010). doi:10.1118/1.3397813

14. Kubicek, J., Penhaker, M.: Fuzzy algorithm for segmentation of images in extraction of objects from MRI. In: International Conference on Advances in Computing, Communications and Informatics (ICACCI, 2014), pp. 1422–1427

15. Kubicek, J., Penhaker, M., Bryjova, I., Kodaj, M.: Articular cartilage defect detection based on image segmentation with colour mapping. In: Lecture Notes in Computer Science (including subseries Lecture Notes in Artificial Intelligence and Lecture Notes in Bioinformatics), vol. 8733, pp. 214–222. Springer, Berlin (2014)

# Segmentation of Macular Lesions Using Active Shape Contour Method

Jan Kubicek, Iveta Bryjova, Marek Penhaker, Jana Javurkova and Lukas Kolarcik

**Abstract** Age-related macular degeneration (ARMD) is one of the most wide-spread diseases of the eye fundus and is the most common cause of vision loss for those over the age of 60. There are several ways to diagnose ARMD. One of them is the Fundus Autofluorescence (FAF) method, and is one of the modalities of Heidelberg Engineering diagnostic devices. The BluePeak™ modality utilizes the fluorescence of lipofuscin (a pigment in the affected cells) to display the extent of the disease's progression. The native image is further evaluated to more precisely specify the diagnosis of the disease—it is necessary to determine the size of the macular lesion area. Calculations of the geometric parameters of macular lesions were conducted in the MATLAB® software; the size of the lesion area was determined using the Image Processing Toolbox. The automated lesion detection method occurs using a parametric active contour (active contours driven by local Gaussian distribution fitting energy) that encloses the affected macular lesion, thereby allowing a precise determination of the affected area. This method is relatively quick for use in clinical practice and allows evaluation the macular lesions exactly based on the proportion with the feature extraction in advance. The proposed methodology is fully automatic. In the algorithm input we define area of interest and initial circle, which is placed inside of the object. Image background is

J. Kubicek · I. Bryjova · M. Penhaker (✉) · J. Javurkova
The Department of Cybernetics and Biomedical Engineering, FEI VSB-TU Ostrava,
17. Listopadu 15, 708 33 Ostrava-Poruba, Czech Republic
e-mail: marek.penhaker@vsb.cz

J. Kubicek
e-mail: jan.kubicek@vsb.cz

I. Bryjova
e-mail: iveta.bryjova@vsb.cz

J. Javurkova
e-mail: javurkova.jana@gmail.com

L. Kolarcik
Clinic of Ophthalmology, University Hospital Ostrava, Ostrava-Poruba, Czech Republic
e-mail: lukas.kolarcik@fno.cz

© Springer International Publishing Switzerland 2016
V. Styskala et al. (eds.), *Intelligent Systems for Computer Modelling*,
Advances in Intelligent Systems and Computing 423,
DOI 10.1007/978-3-319-27644-1_20

213

suppressed by low pass filter. Final contour is formed in consecutive steps, up to shape of macular lesion.

**Keywords** Age-related macular degeneration · Optical coherence tomography · Blue peak™ · Image processing · Active contour · Medical image segmentation

# 1   Introduction

The treatment of many retina diseases adheres to the results of monitoring disease parameters and their temporal evolution. Correct timing of treatment enables better capture of the moment when the patient has the greatest benefit from the selected treatment (hazard ratio, patient stress and expected benefits). Monitoring of the disease evolution also enables appraisal of the evaluation criteria for the purpose of indication and selection of the appropriate type of treatment. One of the determining factors is the extent of the lesion. The evaluation of this parameter is inherently subjective—our aim was finding a suitable method for automatic detection the lesion border line and processing the data into a form that could objectively present the actual state and trends of temporal evolution. Such a procedure can be applied in several areas in ophthalmology. For the design of image processing algorithms and reliability verification, we chose processing of retina image obtained using Spectral is OCT and BluePeak modalities in patients with retinal geographic atrophy resulted from the non-proliferative form of age-related macular degeneration. Another goal will be to verify the solutions in clinical practice, with collateral evaluation of other types of retinal pathology [1–3].

# 2   Age-Related Macular Degeneration

Age-related macular degeneration (ARMD) is one of the most often reason of blindness in developed countries. The disease occurs in 10 % of the population over 60 years and 25 % of the population older than 75 years [1]. The place of ARMD inception is the central part of the retina (macula lutea). The macula is one of the topographical structures of the retina located lateral from the optic nerve (papilla). Accurate evaluation of diagnostic imaging methods is of fundamental importance in determining the course of treatment. ARMD affects the macula in two forms—dry and wet [1]. The dry form of ARMD—nonexudative—is the more common form with a longer progression that may take several months [2, 3]. During the first stage, the disease can develop asymptomatically until the time that metamorphopsias begin to appear. With increasing age, waste substances (lipofuscin) accumulate in the pigment epithelium, leading to deterioration of its functionality and the subsequent atrophy of the affected cells [4]. Prior to the terminal stage of the disease's dry form, the accumulation of these waste substances results in the formation of drusen, i.e. small yellowish deposits that are one of the first risk factors because

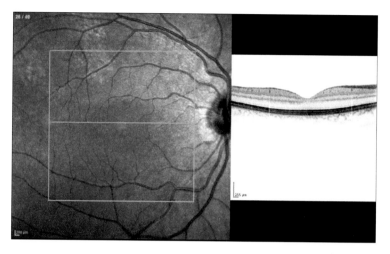

**Fig. 1** Cut of area physiological macula with using infrared shootings (*left*) and OCT cut (*right*)

they cannot be removed or treated in Fig. 1. The wet form—exudative—is characterized by the emergence of a choroidal neovascular membrane (CNVM) in Fig. 2. The disease mainly affects the choroid, inducing the formation of new blood vessels that cause ablation of the retinal layers [5, 6]. Metamorphopsias arise in the early stages; subsequently, newly formed vessels start to bleed, which stimulates the formation of fibrous tissue. The newly created fibrous formation is called a pseudotumour or a disciform scar. This form is far more dangerous than the dry form and has a faster progression [4, 5]. If treatment is not started in time, loss of vision may occur in almost 80 % of the patients in just several months.

## 3   Optical Coherence Tomography (OCT)

OCT is a laser device that employs a superluminiscent diode as the source of its co-herent beam. This diode emits a beam of a suitable wavelength into the eye structure; the beam is reflected by various layers of the retina and interferes with a second reference beam in a detector. The resulting interference signal is then digitized and used to obtain the final image. For developing a method for automatically calculating macular lesions, we used images acquired by a Spectralis unit from Heidelberg Engineering [7–10].

### 3.1   Blue Laser Auto Fluorescence

One modality of OCT devices is BluePeakTM, which excites the ocular fundus (fun-dus oculi) with a blue spectrum laser beam ($\lambda$ = 488 nm). This modality is most com-monly used for diagnosing macular degeneration due to the presence of

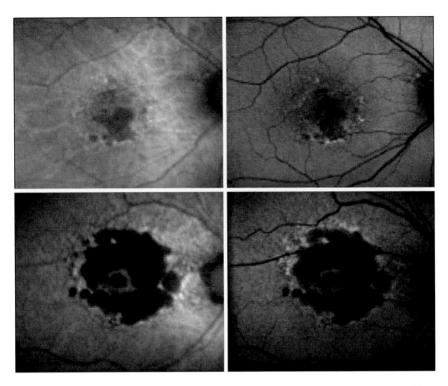

**Fig. 2** The comparison of images, acquired by modified fundus camera (*left*) and confocal laser ophthalmoscopy (*right*)

lipofuscin, which is a specific source of fluorescence. During the examination, the beam is shot into the patient's eye where it induces lipofuscin fluorescence that is caught by the detector and subsequently analysed. The main advantage is the possibility of com-parison with OCT results that provide information on the morphological changes in the retinal pigment epithelium (RPE), while BluePeak$^{TM}$ shows metabolic changes [11–15].

## 4 Proposed Segmentation Software for Macular Lesions Extraction

The main aim of analysis ophthalmologic images is automatic segmentation of areas macular lesions. The essential goal of segmentation procedure is automatic extraction those lesions. Segmentation algorithm is based on active contour method. Defined contour is in iterative steps adjusted to analyzed object. Segmentation output is formed by smooth and closed curve which reflects area, where is placed analyzed object. After loading input image data is needed to select RoI (Region of

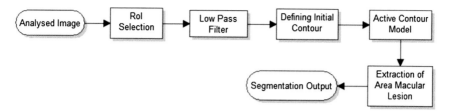

**Fig. 3** The block diagram of segmentation method for retinal lesions extraction

Interest). Low pass filter is used for suppression image background. The center of initial circle is consequently selected. After taking 100 iterations final contour is formed. This contour has to reflect analyzed macular lesion (Fig. 3).

## 4.1 Defining of Geometrical Contour

Used type of geometrical contour is derived from implicit equation of initial curve. For purposes of our analyses circle with zero shift is used:

$$x^2 + y^2 = r^2 \rightarrow x^2 + y^2 - r^2 = 0 \tag{1}$$

Initialization function is formed in individual iteration steps to shape of macular lesion. On level set method, analyzed image is divided by contour into inside and outside part. Image is composed from three parts: Inside area is consist by negative values (negative values of shortest Euclidean distance of points from contour). Inside area (positive value of the shortest Euclidean distance points from contour) and contour with zero value. The further away from the point of the curve, the larger the resulting value. The result of this procedure is the cone which defines the distance the positive and negative values from zero, and thereby to form level set area. Level set area is defined by level set function Φ which is given by following equation:

$$\Phi(x, y, t) \tag{2}$$

Value range of function Φ is placed to $\mathbb{R}^3$. It is not just curve, but whole domain is defined which is consecutive changed by the time. For the simplest case for defining level set evolution it is necessary to define gradient. Gradient is needed for performing proper evolution of level set area:

$$\nabla\Phi = (\frac{\partial\Phi}{\partial x}, \frac{\partial\Phi}{\partial t}) \tag{3}$$

Evolution of level set by the time is given by partial derivation level set function $\Phi$ by the time $\frac{\partial \Phi}{\partial t}$.

Final contour is formed by the time in the direction of normal when velocity of evolution "c" is being stable.

$$\frac{\partial \Phi}{\partial t} = c.N \tag{4}$$

Normal vector is given by relationship:

$$N = \frac{\nabla \Phi}{|\nabla \Phi|} \tag{5}$$

Direction of normal is determined by gradient. Normal vector is vertical to tangent of contour which is developed in direction of this normal multiplied by constant. Instead of constant velocity we can use function divergence of normal. Divergence function determines when vectors converge to some particular point. In the first step normal and divergences are calculated. Negative divergence denotes on convergence of vectors. Evolution of level set area is performed by this approach. This level set area is independent on image values.

The most frequently used contours are based on principle of minimization of energy functional. Functional is representation which assigns real number. Contour is placed in functional and its size of energy is controlled. If the energy is too large, the contour is gradually deformed into a shape that reduces energy. This procedure is performed until we reach energy minimization. Active contour method is iterative algorithm, which forms final contour in consecutive steps. The key parameter is number of those steps. We must keep perimeter of analyzed object. For our purpose 100 iterations have been used for reaching shape of macular lesions [13].

## 5   Data Analysis and Segmentation Results

For testing of the designed software, 40 patient's records of macular lesions have been used. Testing has been performed with same requirement and its background suppression and reaching of final contour macular lesion. Segmentation gave satisfactory results for 35 patients. On the rest images we had problems with adjacent blood vessels which we are not able to suppress. This fact causes worse effectivity of detection. Images obtained through the BluePeak modality provide unique information about the condition of macular degeneration and its terminal stage—geographic atrophy. Thanks to these high contrast images, it is possible to determine the geometric parameters of macular lesions. For the subsequent processing of native images, we used the MATLAB® interactive programming environment. The aim was to quantify the area of the ocular fundus of macular lesions. This geometric parameter allows clinicians to clarify and predict the further development of the

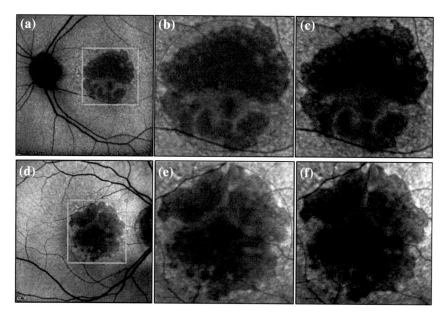

**Fig. 4** Selection of macular lesion area (**a, d**), the contour's initial state before segmentation process (**b, e**) and resulting segmentation outputs (**c, f**)

disease. Emphasis was particularly placed on those segmentation methods that allow for an automated analysis. The geometric active contours driven by local Gaussian distribution fitting energy method—one of the level-set segmentation procedures—proved to be the best. The algorithm is divided into several basic parts. First, the patient's image is loaded into the device and the macular lesion zone is framed; calculation of the macular lesion area follows. The selection of the macular lesion area (Fig. 4) is indicated using the active contour model (the gradual shaping of contours up to the edge of the object in the image). With zero approximation, the contour is defined as initial circle whose size is consecutively adapted to the size of the lesion. After the contour completes the segmentation, the number of pixels contained therein is calculated. Here we use the properties of level-set methods that divide the image into two parts: the part inside the contour and the part outside the contour. The contour corresponds to zero values. The sum of the number of pixels is defined by a cycle that evaluates all the pixels in the contour as being logical ones. The conversion of the number of pixels per unit area occurs via image resolution. In Heidelberg Spectralis OCT devices, this resolution corresponds to 200 μm. The resulting geometric parameter, i.e. the size of lesion area, is $S = 7.156$ mm$^2$ in the first case (in Fig. 4c) and $S = 8.9796$ mm$^2$ in the latter case (in Fig. 4f). A current shortfall of this analysis is the vascular bed, which is one of the most contrasted parts of the ocular fundus and the active contour extends behind it. Further development of this method should focus on subtracting the area of the vascular bed, which will significantly improve the diagnosis of retinal disease.

# 6  Conclusion

Our research is focused on software solution for fully automatic macular lesions extraction. 40 patients records have been used for purposes of our analysis. The proposed segmentation software is fully automatic. The defining RoI and putting the center of initial curve are manual procedures. After taking those procedures, final contour is performed. The key parameter is number of iterative steps of active contour. For our analysis 100 iterative steps have been used. From the statistical view, the segmentation software reached satisfactory results for 87 % images. The rest of them contain blood vessel's system. The problem is that blood vessels are represented by same color spectrum as analyzed macular lesions and therefore it is quite complicated to suppress those structures. This analysis is being solved with cooperation University hospital of Ostrava. Proposed software method is currently being tested in clinical practice. Automatic evaluation of geometrical parameters in ophthalmologic records in clinical practice is really needed. It allows following up progression of lesions within the time. Because late performed diagnosis causes blindness our research is focused primarily to this area. Thanks to early and precise diagnosis of disease it is possible to begin treatment which is applied in the form of appropriate medicaments in the case of wet form. In the coming time it would be useful to extend our application for using on wet and dry form of macular lesions. Analysis of some images with geographic atrophy has been unsuccessful. This fact has been caused by badly contrast of papilla and highlighted blood vessels. These blood vessels have been kept in the image and disturbed of expansion active contour. Active contour is not able to keep only into the analyzed object but continues out of area of interest. We are currently working on more sophisticated algorithm which should be able to suppress blood vessels and keep just macular lesions.

**Acknowledgements** The work and the contributions were supported by the project SP2015/179 'Biomedicínské inženýrské systémy XI' and This work is partially supported by the Science and Research Fund 2014 of the Moravia-Silesian Region, Czech Republic and this paper has been elaborated in the framework of the project "Support research and development in the Moravian-Silesian Region 2014 DT 1—Research Teams" (RRC/07/2014). Financed from the budget of the Moravian-Silesian Region.

# References

1. Wang, Z.L., et al.: Bevacizumab cured age-related macular degeneration (AMD) via down-regulate TLR2 pathway. Central Eur. J. Biol. **9**(5), 469–475 (2014). doi:10.2478/s11535-014-0290-5
2. Christen, W.G., Chew, E.Y.: Does long-term aspirin use increase the risk of neovascular age-related macular degeneration?. Expert Opin. Drug Saf. **13**(4), 421–429 (2014). doi:10.1517/14740338.2014.889680.

3. Pustkova, R., et al.: Measurement and calculation of cerebrospinal fluid in proportion to the skull. In: 2010 9th Roedunet International Conference (RoEduNet) (2010)
4. Cheung, L.K., Eaton, A.: Age-related macular degeneration. Pharmacotherapy: J. Human Pharmacol. Drug Ther. **33**(8), 838–855 (2013). doi:10.1002/phar.1264
5. Penhaker, M., Matejka, V.: Image registration in neurology applications. In: 2010 International Conference on Networking and Information Technology (ICNIT) (2010)
6. Tsika, Ch., Tsilimbaris, M.K., Makridaki, M., Kontadakis, G., Plainis, S., Mos-chandreas, J.: Assessment of macular pigment optical density (MPOD) in patients with unilateral wet age-related macular degeneration (AMD). Acta Ophthalmol. **89**(7), e573–e578 (2011)
7. Stetson, P.F., et al.: OCT minimum intensity as a predictor of geographic atrophy enlargement. Invest. Ophthalmol. **55**(2), 792–800 (2014). doi:10.1167/iovs.13-13199
8. Alam, S., et al.: Clinical Application of rapid serial fourier-domain optical coherence tomography for macular imaging. Ophthalmology **113**(8), 1425–1431 (2006). doi:10.1016/j.ophtha.2006.03.020
9. Kubicek, J., et al.: Segmentation of MRI data to extract the blood vessels based on fuzzy thresholding. In: New Trends in Intelligent Information and Database Systems, pp. 43–52. Springer International Publishing, Berlin (2015)
10. Coscas, G., et al.: Optical coherence tomography in age-related macular degener-ation: OCT in AMD. Springer, Heidelberg (2009). ISBN 978-364-2014-680
11. Besirli, C.G., Comer, G.M.: High-resolution OCT imaging of RPE degeneration in bilateral diffuse uveal melanocytic proliferation. Ophthalmic Surg. Lasers Imaging. **41**(6), S96–S100 (2010). doi:10.3928/15428877-20101031-03
12. Blue Laser Autofluorescence. A supplement to Ophthalmology Times Europe: Blue laser autofluorescence [online]. Advanstar Communications, Chester (2009) [cit. 2013-12-03]. ISSN 1753-3066)
13. Wang, L., et al.: Active contour driven by local Gaussian distribution fitting energy. Sig. Proc. **89**(12), 2435–2447 (2009). doi:10.1016/j.sigpro
14. Kubicek, J., Penhaker, M.: Fuzzy algorithm for segmentation of images in extraction of objects from MRI. In: 2014 International Conference on Advances in Computing, Communications and Informatics (ICACCI), IEEE (2014)
15. Kubicek, J., et al.: Articular cartilage defect detection based on image segmentation with colour mapping, in computational collective intelligence. Technologies and Applications, pp. 214–222. Springer International Publishing, Berlin (2014)

# Mathematical and Experimental Analyses of a Compression Ignition (CI) Engine Run on a Bioethanol Diesel Emulsion

**Murugan Sivalingam, Dulari Hasdah and Bohumil Horák**

**Abstract** In this study, a possibility of using a bioethanol-diesel emulsion as an alternative fuel for compression ignition (CI) engines is explored. For the investigation bioethanol was obtained from Madhuca Indica flower by fermentation process. Two important emission parameters nitric oxide (NO) and smoke emissions from the emulsion fueled single cylinder, four stroke, air cooled direct injection (DI) CI engine were determined and compared with those of diesel operation at different load conditions. The experimental results were also validated by a mathematical modelling using a MATLAB program and are presented in this paper.

**Keywords** Bioethanol · Diesel engine · Modelling · MATLAB

## 1 Introduction

Ethanol derived from biomass sources is known as bioethanol. The utilization of bioethanol in internal combustion (IC) engines is paid more attention because biomass is renewable and abundantly available [1–3]. Numerous research works have been documented to use bioethanol in the form of blending/emulsion, fumigation, dual injection, surface ignition etc. in CI engines [4]. In recent years, the study and control of emissions from CI engines have been highly concentrated.

M. Sivalingam (✉)
Centre ENET, Ostrava, The Czech Republic
e-mail: muruganresearch@yahoo.com

B. Horák (✉)
Department of Cybernetics and Biomedical Engineering, VSB Technical University,
Ostrava, The Czech Republic
e-mail: bohumil.horak@vsb.cz

M. Sivalingam · D. Hasdah
Department of Mechanical Engineering, National Institute of Technology,
Rourkela, India

© Springer International Publishing Switzerland 2016
V. Styskala et al. (eds.), *Intelligent Systems for Computer Modelling*,
Advances in Intelligent Systems and Computing 423,
DOI 10.1007/978-3-319-27644-1_21

223

Researchers suggested that theoretical analysis accomplished by mathematical modelling or numerical solutions using computer program or computational fluid dynamics (CFD) can give more fruitful predictions for the engine parameters [5–7]. They used single zone or multi zone modelling to validate the experimental results obtained for different combustion, performance and emission parameters. In this study, a mathematical modelling was developed using a MATLAB program to validate the experimental results obtained from a single cylinder, four stroke, air cooled, DI diesel engine was run on the BMDE15 emulsion. The BMDE15 refers to an emulsion containing 84 % bioethanol, 15 % diesel and 1 % surfactant in it. A MATLAB program was developed for a two zone model for the validation. One zone consisted of pure air called the non-burning zone, and the other consisted of fuel and combustion products, called the burning zone. The combustion parameters, such as ignition delay and heat release rate and the chemical equilibrium composition were calculated theoretically, using the two zone model. With the help of these parameters as inputs, the NO and soot emissions were determined. The experimental results were validated with the help of the theoretical results and presented in this paper.

## 2   Materials and Method

In this experimental investigation, bioethanol was emulsified with diesel (BMDE15) used as an alternative fuel in a single cylinder, four stroke direct injection (DI) diesel engine. The procedure of producing the bioethanol has already been described in [8]. The engine was coupled to an electrical dynamometer to provide the brake load with an electric panel. A fuel control valve switch fuels, fuel sensor measure consumption, air consumption was measured with the help air sensor on the air box. The collected data were displayed on the monitor of the computer. For the emission measurements, an exhaust gas analyzer was used to measure the level of HC, $CO_2$, CO, and NO. A diesel smoke meter was used to measure the smoke in the engine exhaust. Initially, the engine was operated with diesel for obtaining the reference data. The experimental set up used in this investigation is shown in Fig. 1.

The physico chemical properties of the BMDE15 emulsion are tabulated in Table 1 in comparison with those of diesel.

## 3   Mathematical Modelling

In a CI engine, the fuel air mixture is obtained inside the combustion chamber of the engine. The injected fuel absorbs the heat from the surrounding air and vaporizes. Further, the fuel vapor mixes with the available air in the cylinder. Depending on the spray, the fuel air mixture is obtained in the cylinder. A computer program using

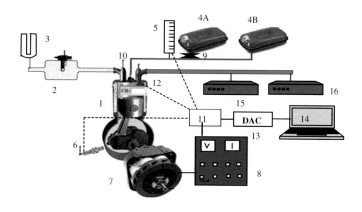

1.Engine              5. Burette              10. Fuel injector          15. Smoke meter
2.Air box             6. Pressure transducer  11. Control panel          16. Gas analyser
3. U-tube manometer   7.Alternator            12. Pressure transducer

4A.Diesel tank        8. Load cell            13. Data acquisition
4B.Emulsion tank      9. Valve                14. Personal computer

**Fig. 1** Experimental set up

**Table 1** Properties of diesel and BMDE15 [8]

| Description | Diesel | BMDE15 |
|---|---|---|
| Chemical formula | $C_{16}H_{34}$ | $C_{5.471}H_{6.039}O$ |
| Molecular weight | 170 | 48 |
| Density at 40 °C | 2.4 | 1.73 |
| Carbon | 86 | 65.65 |
| Hydrogen | 13.60 | 10.21 |
| Nitrogen | 0.18 | 0.14 |
| Sulfur | 0.22 | 0.01 |
| Oxygen by difference | 0 | 24 |

MATLAB was generated, with all the above mentioned equations and considering all the values of the constants, in order to predict the NO and soot emissions from the engine run on the BMDE15 and diesel fuel.

## 3.1 Energy Equations

During the compression stroke, only one zone (of pure air) exists. Then, the first law of thermodynamics for a closed system is applied, together with the perfect gas state equation. The change in internal energy is expressed [9] as follows:

$$\frac{d(mu)}{d\theta} = \frac{dQ_r}{d\theta} - \frac{dQ_h}{d\theta} - \frac{dW}{d\theta} \qquad (1)$$

By replacing the work transfer term $dW/d\theta$ with $PdV/d\theta$ or by the ideal gas law $PV = mRT$, the above equation can be rearranged as,

$$m\frac{du}{d\theta} = \frac{dQ_r}{d\theta} - hA\frac{dT}{d\theta} - RT\frac{dV}{d\theta} \qquad (2)$$

where, V is the instantaneous cylinder volume with respect to the crank angle, which is given by,

$$V = V_{cl} + \left(\pi D^2/4\right)r\left[1 + \lambda^{-1} - \cos\varphi - \left(\lambda^{-2} - \sin^2\varphi\right)^{1/2}\right] \qquad (3)$$

In the above equations, the term dQ is given as the fourth order polynomial expression of the absolute temperature T, including the enthalpy of formation at absolute zero. The internal energy calculation as a function of temperature is:

$$\frac{hi}{R_{mol}T} = ai1 + ai2/2T + ai3/3T^2 + ai4/4T^3 + ai5/5T^4 + ai6/3T^5 \qquad (4)$$

$$ui = hi - RT \qquad (5)$$

For the surrounding air zone, which only loses the mass (air) to the burning zone, the first law of thermodynamics for the unburned zone is written as,

$$dE = dQ - pdV - h_a dm_a \qquad (6)$$

The burning zone not only receives the mass from the air zone, but also there is an enthalpy flow from the fuel, which is ready to be burned in the time step. So, the first law of thermodynamics for the burning zone becomes

$$dE = dQ - pdV + h_a dm_a + h_f dm_f \qquad (7)$$

The first law of thermodynamics for the combustion in time step dt is

$$f(E) = E(T_2) - E(T_1) - dQ + dW + dm_f Q_{vs} = 0 \qquad (8)$$

If $f(E)$ is greater than the accuracy, the required new value of $T_2$ is calculated using the Newton-Raphson numerical method. The unburned zone temperature is calculated using the equation,

$$T_u = T_{soc}\left(\frac{P}{P_{soc}}\right)^{\gamma - 1/\gamma} \qquad (9)$$

## 3.2 Heat Transfer Model

The heat transfer between the cylinder trapped mass and the surrounding walls is calculated, using the formula of Anand [10] which is as follows;

$$dQ/dt = a\frac{\lambda_g}{D}(Re)^b(T_w - T_g) + c\cdot\left(T_w^4 - T_g^4\right) \tag{10}$$

In this equation 'Tw' is the cylinder wall temperature which is assumed as 450 K, and a, b, c are constants. The constant values are taken as, $a = 0.2626$, $b = 0.6$, $c = 5.67 \times 10^{-8}$ W/m$^2$/K.

## 3.3 Ignition Delay

The time delay between the start of injection and the start of combustion is defined as the ignition delay period. In the combustion model, the ignition delay is also taken into account. The ignition delay period is calculated by integrating Wolfer's relation, using the trapezoidal rule [11].

$$\int_{t_{inj}}^{t_{ign}} \frac{dt}{t(p,T)} = \frac{1}{K_{t_{inj}}}\int_{t_{inj}}^{t_{ign}} \frac{dt}{(p(t))^{-q}\exp(\frac{E}{RT(t)})} = 1 \tag{11}$$

The values of various constants corresponding to a DI diesel engine are K = 2272; q = −1.19; E/R = 4650.
where
K     thermal conductivity
q     heat losses
E/R   activation energy/universal gas constant.

The Wiebe functions [12] for the non-dimensional burn fraction x as a function of the degrees of crank angle can be written as

$$x = 1 - \exp\left[-6.908\left(\frac{\theta - \theta o}{\Delta\theta}\right)^{m+1}\right] \tag{12}$$

The heat release rate calculated with the help of the Wiebe function is,

$$\frac{dQ_c}{d\theta} = 6.908(m+1)\left(\frac{Q_{av}}{\Delta\theta}\right)\left(\frac{\theta - \theta_o}{\Delta\theta}\right)^m \exp\left[-6.908\left(\frac{\theta - \theta o}{\Delta\theta}\right)^{m+1}\right] \tag{13}$$

where x is the mass fraction burned, $\theta o$ is the start of combustion and $\Delta\theta$ is the combustion duration. The parameter m represents the rate of combustion. $Q_{av}$ is the heat released per cycle. The value of m for both the fuels is taken as 3.0.

## 3.4  Nitric Oxide (NO) Formation Model

The current approach to model the $NO_x$ emissions from diesel engines is, to use the extended Zeldovich thermal NO mechanism, by neglecting other sources of $NO_x$ formation. The extended Zeldovich mechanism consists of the following reactions,

$$O + N_2 \rightarrow NO + N \tag{14}$$

$$N + OH \rightarrow NO + H \tag{15}$$

This mechanism can be written as an explicit expression for the rate of change of the concentration of NO.

The change of NO concentration is expressed as follows:

$$(d(NO))/dt = 2\left(1 - \alpha^2\right) R_1/(1 + \alpha R_1/(R_2 + R_3)) \tag{16}$$

where $R_i$ is the one-way equilibrium rate for the reaction i, defined as,

$$R_1 = k_{1f}(N)e(NO)e, \quad R_2 = k_{2f}(N)e(O_2)e, \tag{17}$$

$$R_3 = k_{3f}(N)e(OH)e, \quad \alpha = (NO)/(NO)e \tag{18}$$

## 3.5  The Net Soot Formation Model

The exhaust of the CI engine contains solid carbon soot particles that are generated in the fuel rich regions inside the cylinder during combustion. The net soot formation rate was calculated by using the semi-empirical model. According to this model, the soot formation rate (index sf) and soot oxidation rate (index sc) were given by,

$$\frac{dm_{sf}}{dt} = A_{sf}\left(m_{f.ev} - m_{f.bu}\right)^{0.8} p^{0.5} \exp\left(-E_{sf}/R_{mol}T\right) \tag{19}$$

$$\frac{dm_{sf}}{dt} = A_{sc}m_{sn}(p_{o2}/p)p^{1.8} \exp\left(-E_{sc}/R_{mol}T\right) \tag{20}$$

where, the pressures are expressed in bar and $d_{mf}$ is the unburned fuel mass in kg to be burned in time step dt. Therefore, the net soot formation rate is expressed as;

$$\frac{dm_{sn}}{dt} = \frac{dm_{sf}}{dt} - \frac{dm_{sc}}{dt} \qquad (21)$$

## 4  Results and Discussion

In a CI engine, the $NO_x$ emission is one of the major pollutants and is predominantly influenced by the amount of oxygen available, and the in-cylinder temperature [13]. The comparison between the simulated and experimental results is shown in Fig. 2. The brake specific NO emissions are obtained from the simulation and experiments show a declining trend as the load increases. The NO emission values obtained from the simulation and experiments are found to be lower than those of diesel operation, because of the higher latent heat of the vaporization of BMDE15. An overall marginal deviation less than 4 % is noticed between the simulation and experimental results. This may be due to experimental errors while reading measurements.

The variation of the simulated and experimental results of smoke emission for diesel and BMDE15 is shown in Fig. 3. The simulated results of diesel for the smoke emission are found to be high compared to the simulated and experimental results. The smoke emission is a result of the oxygen unavailability in the diffusion combustion phase, use of high molecular weight fuel and the aromatic content of fuel. Diesel has a high carbon to hydrogen ratio, high molecular weight, less oxygen and high aromatic content. Hence, higher smoke emission is observed with the diesel operation. The deviation between the simulated and experimental values is about 3 and 4 %.

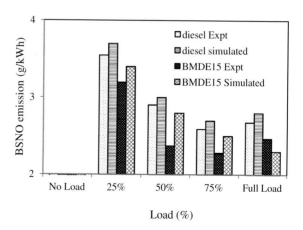

**Fig. 2** BSNO emission with load

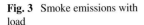

Fig. 3 Smoke emissions with load

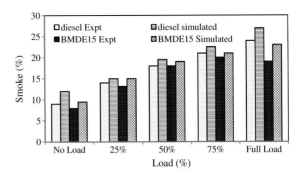

## 5 Conclusions

A comprehensive two zone model was developed to validate the experimental results obtained from a single cylinder, four stroke, air cooled, DI diesel engine run on two different fuels, viz., diesel and BMDE15. The following is the summary of the results;

(i) The NO emission of the engine run on BMDE15 is found to be lower than that of diesel which is due to the high latent heat of vaporisation. This is validated with the NO emission model. The deviation between the simulated and experimental results in the diesel and BMDE15 operations are about 7.1 and 5 % respectively, at full load.

(ii) The smoke opacity is found to be lower for the BMDE15 operation compared to that of diesel. The smoke opacity of BMDE15 in the simulation condition is higher by about 4 % than that of the experiment.

**Acknowledgements** This paper has been elaborated in the framework of the project New creative teams in priorities of scientific research, reg. no. CZ.1.07/2.3.00/30.0055, supported by Operational Programme Education for Competitiveness and co-financed by the European Social Fund and the state budget of the Czech Republic. The authors also thank National Institute of Technology, Rourkela for carrying out the experiments.

## References

1. Leite, R.C.C., Leal, M.R.L.V., Cortez, L.A.B., Barbosa, L.A., Griffin, W.M., Scandiffio, M.I. G.: Can Brazil replace 5 % of the 2025 gasoline world demand with ethanol? Energy **34**, 655–661 (2009)
2. Demirbas, A.: Bioethanol from cellulosic materials: a renewable motor fuel from biomass. Energy Sour. **27**, 327–337 (2005)
3. Hahn-Hägerdal, B., Galbe, M., Gorwa-Grauslund, M.F., Lidén, G., Zacchi, G.: Bio-ethanol-the fuel of tomorrow from the residues of today. Trends Biotechnol. **24**, 549–556 (2006)

4. Ecklund, E.E., Bechtold, R.L., Timbario, T.J., McCallum, P.W.: State-of-the art report on the use of alcohols in diesel engines. SAE Technical Paper 840118, pp. 1684–1702 (1984)
5. Reitz, R.D., Bracco, F.V.: On the dependence of spray angle and other spray parameters on nozzle design and operating conditions. SAE technical paper no. 790494 (1979)
6. Patil, S.: Thermodynamic modelling for performance analysis of compression ignition engine fuelled with biodiesel and its blends with diesel. Int. J. Recent Technol. Eng. (IJRTE) 1, 134–138 (2013)
7. Awad, S., Varuvel, E.G., Loubar, K., Tazerout, M.: Single zone combustion modeling of biodiesel from wastes in a diesel engine. Fuel 106, 558–568 (2013)
8. Hansdah, D., Murugan, S., Das, L.M.: Experimental studies on a DI diesel engine fueled with bioethanol-diesel emulsions. Alexandria Eng. J. 52, 267–276 (2013)
9. Ramadhas, A.S., Jayaraj, S., Muraleedharan, C.: Theoretical modeling and experimental studies on biodiesel-fueled engine—technical note. Renew. Energy 31, 1813–1826 (2006)
10. Annand, W.J.D.: Heat transfer in the cylinders of reciprocating internal combustion engines. Proc. Inst. Mech. Eng. 177, 973–990 (1963)
11. Rakopoulos, C.D., Rakopoulos, D.C., Giakoumis, E.G., Kyritsis, D.C.: Validation and sensitivity analysis of a two-zone diesel engine model for combustion and emissions prediction. Energy Convers. Manag. 45, 1471–1479 (2004)
12. Gogoi, T.K., Baruah, D.C.: A cycle simulation model for predicting the performance of a diesel engine fuelled by diesel and biodiesel blends. Energy 35, 1317–1323 (2010)
13. Heywood, J.B.: Internal Combustion Engines Fundamentals, pp. 491–667. McGraw Hill Publications, New York (1988)

# Modelling of Antiphase Noise Reduction of a Transformer

Stanislav Misak, Viktor Pokorny and Petr Kacor

**Abstract** This article discusses about development of research of using antiphase elimination of noise reduction. It dedicated to modeling of acoustic phenomena by software AnSys and possibility of simulation and modeling of acoustic short-circuit over the antiphase elimination. The team of authors of this article have been involved in the issue of eliminating noise emissions by electronic equipment from the year 2009. This theme is extremely relevant at present since the health of the human population is a definite priority in human efforts. The first functional prototype for an Antiphase Eliminator, which was installed on a power transformer, was successfully constructed in 2012 on the basis of research. The team develops anti phase eliminator for over 6 years. Now after completion of analog prototype began with digitization and thus building another prototype. This new prototype will processes the signals into digital form. Because the efficiency of the analog prototype is relatively low, the team began together with the digitization also work on computer models that would reveal possible causes of low efficiency. These models already subtracted signals only, as it was on the start of research. These models simulate real acoustic propagation in space, his reflections and the resulting impact of the action of antiphase eliminator.

**Keywords** Anti-phase · Noise · Acoustic signal · Modeling · Ansys · Boundary conditions

S. Misak (✉) · P. Kacor
Faculty of Electrical Engineering and Computer Science,
Department of Electrical Power Engineering, VŠB-Technical
University of Ostrava, Ostrava, Czech Republic
e-mail: stanislav.misak@vsb.cz

P. Kacor
e-mail: petr.kacor@vsb.cz

V. Pokorny (✉)
Faculty of Electrical Engineering and Computer Science, VŠB-Technical
University of Ostrava, Research Programme 5, Ostrava, Czech Republic
e-mail: viktor.pokorny@vsb.cz

© Springer International Publishing Switzerland 2016
V. Styskala et al. (eds.), *Intelligent Systems for Computer Modelling*,
Advances in Intelligent Systems and Computing 423,
DOI 10.1007/978-3-319-27644-1_22

# 1  Introduction

Currently, the health of human population in any activity is a priority of human pursuit. Indivisible factor having an adverse impact on the human organism is a parasitic noise around us. Human physiology is not adapted to deal with noise emissions, which currently man produces, therefore are issued orders, regulations and directives governing the noise level to human in various activities in various areas. The aim of designers is the parasitic noise as possible to eliminate [1].

# 2  Antiphase

To understand the system is important to know basic, namely waves alone. Phase of the wave is a dimensionless quantity that determines the relation variable waves, (e.g. displacement noise) at that place and time and to the state variables characteristic waves in temporal and spatial origin. Dependence characteristic of variables determines the shape of "waves" regardless of its dissemination. Phase is a parameter, which depends on the timing characteristic values in a fixed location, which the wave passes, respectively spatial field characteristic values for a fixed moment in time. Noise so we can capture by the curve demonstrate displacement in time. Anti-phase is turning the current signal of 180°.

The result of the exact anti-phase is an absolute deduction of both signals and therefore their complete elimination (Fig. 1) [1, 2].

The first experiments with an acoustic antiphase wave began to be carried out at VŠB-Technical University of Ostrava in 2008. This impulse arose after measuring the noise of wind power plants as a result of people in their surroundings complaining about the noise. The commissioner of the measuring required values after the analysis as well as a plan for improving the current problematic state of affairs. The idea of making use of an antiphase wave consequently came about as a way of actively counteracting this parasitic noise.

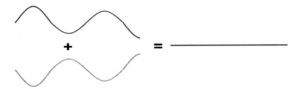

**Fig. 1**  Generating of antiphase wave

# 3 Results of the Antiphase Eliminator Model

The polar graph compares all three situations, the noise of the room (blue), the noise of the transformer (orange) and noise during elimination by antiphase (violet) (Fig. 2).

The main result of this research is the elimination of the noise of the transformer by about 17 dB.

Based on the results achieved by the model of the Antiphase Eliminator, we began to construct a prototype which could be tested on an actual transformer.

Compared with the original intention, the construction is different as the elimination is carried out only from one side. The reason for this is the complexity of the spreading of the acoustic wave. We know from the measuring on the model that the production of feedback in the system is a serious problem. The effect of two speakers against one other would cause serious problems with this phenomenon. The entire system will be one-sided after the function verification (Fig. 3).

**Fig. 2** Polar graph comparing the results

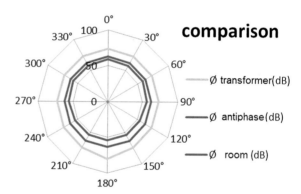

**Fig. 3** Elimination of the transformer noise with the antiphase

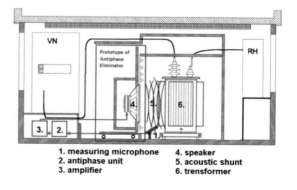

## 4   Results of the Antiphase Eliminator Model

During all connections and measurements, a spectral analysis of transformer noise was continuously performed both in the regular operation and in elimination.

From the graphs of spectral analysis of transformer noise both during elimination and without the elimination, it is clear that the elimination was also successful in a real transformer. Frequency 101 Hz was suppressed by 13.1 dB and also the remaining frequencies ranged below −60 dB.

The graph on Fig. 4 presents the spectral analysis of noise transformer T616; it is clearly evident that the carrier frequency 101 Hz stands above the rest of the spectrum. In addition, the multiples of this frequency of 200, 300 and 500 Hz are evident. All these frequencies exceed level −50 dB (Fig. 5).

All of the achieved results indicate the success of this method in eliminating noise. The actual core of the prototype is currently established as only analogical equipment which is additionally made up of elements which are far from sufficient for the eventual requirements. These elements are unnecessarily uneconomical, both in terms of their purchase price and in terms of their operations.

**Fig. 4** Spectral analysis of the noise of transformer T616

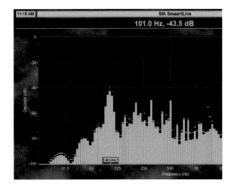

**Fig. 5** Spectral analysis of the noise of transformer T616 during elimination

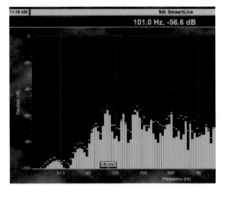

The prototype supplied with the protection of industrial ownership, at this point an Applied Model, was created on the basis of these successful tests. This Applied Model was registered with the Office of Industrial Property under the registration number UPV no. 25699 and further as the functional sample no. 039/22-05-2012_F. There will be efforts carried out in order to obtain patent protection.

A consequent step in this research will be the digitalization and specification of the entire Antiphase Eliminator [1, 3, 4].

## 5 Modeling of Acoustic Short

With the idea to make the whole system modeling came the team together with the idea to continue the development of a prototype and make digitization of the prototype. If the prototype is to be more effective, they must find out exactly how acoustic waves propagate and what occurs when exposed to parasitic antiphase wave acoustic signals from the madding sources [5].

Acoustics is the study of the generation, propagation, absorption, and reflection of sound pressure waves in a fluid medium. Typical quantities of interest are the pressure distribution in the fluid at different frequencies, pressure gradient, particle velocity, the sound pressure level, as well as, scattering, diffraction, transmission, radiation, attenuation, and dispersion of acoustic waves.

In normal gases, at audible frequencies, the pressure fluctuations occur under essentially adiabatic conditions (no heat is transferred between adjacent gas particles). Speed of sound then becomes

$$c = \sqrt{\frac{\gamma P}{\rho}} \tag{1}$$

where $\gamma = Cp/Cv = 1.4$ for air and $P = \rho RT$ (Ideal Gas Law).

## 5.1 Conditions Acoustic Field and Boundary Conditions

In acoustic fluid-structural interaction (FSI) problems, the structural dynamics equation must be considered along with the Navier-Stokes equations of fluid momentum and the flow continuity equation. From the law of conservation of mass law comes the continuity equation:

$$\frac{\partial \rho}{\partial t} + \frac{\partial (\rho Vx)}{\partial x} + \frac{\partial (\rho Vy)}{\partial y} + \frac{\partial (\rho Vz)}{\partial z} = 0 \tag{2}$$

where vx, vy and vz = components of the velocity vector in the x, y and z directions, respectively, $\rho$ = density, x, y, z = global Cartesian coordinates, t = time [5].
The Navier-Stokes equations are as follows:

$$\frac{\partial \rho Vx}{\partial t} + \frac{\partial (\rho VxVx)}{\partial x} + \frac{\partial (\rho VyVx)}{\partial y} + \frac{\partial (\rho VzVx)}{\partial z}$$

$$= \rho gx - \frac{\partial P}{\partial x} + Rx + \frac{\partial}{\partial x} + \left( \mu e \frac{\partial Vx}{\partial x} \right) + \frac{\partial}{\partial y} \left( \mu e \frac{\partial Vx}{\partial y} \right) + \frac{\partial}{\partial z} \left( \mu e \frac{\partial Vx}{\partial z} \right) + Tx$$

(3)

$$\frac{\partial \rho Vy}{\partial t} + \frac{\partial (\rho VxVy)}{\partial x} + \frac{\partial (\rho VxVy)}{\partial y} + \frac{\partial (\rho VxVy)}{\partial z}$$

$$= \rho gy - \frac{\partial P}{\partial y} + Ry + \frac{\partial}{\partial x} + \left( \mu e \frac{\partial Vy}{\partial x} \right) + \frac{\partial}{\partial y} \left( \mu e \frac{\partial Vy}{\partial y} \right) + \frac{\partial}{\partial z} \left( \mu e \frac{\partial Vy}{\partial z} \right) + Ty$$

(4)

$$\frac{\partial \rho Vz}{\partial t} + \frac{\partial (\rho VxVz)}{\partial x} + \frac{\partial (\rho VyVz)}{\partial y} + \frac{\partial (\rho VzVz)}{\partial z}$$

$$= \rho gz - \frac{\partial P}{\partial z} + Rz + \frac{\partial}{\partial x} + \left( \mu e \frac{\partial Vz}{\partial x} \right) + \frac{\partial}{\partial y} \left( \mu e \frac{\partial Vz}{\partial y} \right) + \frac{\partial}{\partial z} \left( \mu e \frac{\partial Vz}{\partial z} \right) + Tz$$

(5)

where gx, gy, gz = components of acceleration due to gravity, $\rho$ = density,
$\mu e$ = effective viscosity, Rx, Ry, Rz = distributed resistances, Tx, Ty, Tz = viscous
loss terms.

The fluid momentum (Navier-Stokes) equations and continuity equations are
simplified to get the acoustic wave equation using the following assumptions:

1. The fluid is compressible (density changes due to pressure variations).
2. There is no mean flow of the fluid [5].

The acoustic wave equation is given by:

$$\nabla \left( \frac{1}{\rho 0} \nabla \rho \right) - \frac{1}{\rho 0 c^2} \frac{\partial^2 \rho}{\partial t^2} + \nabla \left[ \frac{4\mu}{3\rho 0} \nabla \left( \frac{1}{\rho 0 c^2} \frac{\partial \rho}{\partial t} \right) \right]$$

$$= -\frac{\partial}{\partial t} \left( \frac{Q}{\rho 0} \right) + \nabla \left[ \frac{4\mu}{3\rho 0} \nabla \left( \frac{Q}{\rho 0} \right) \right]$$

(6)

where c = speed of sound $\left( \sqrt{K/\rho_0} \right)$ in fluid medium, $\rho$o = mean fluid density,
K = bulk modulus of fluid, $\mu$ = dynamic viscosity, p = acoustic pressure (= p(x, y, z,
t)), Q = mass source in the continuity equation, t = time.

Harmonically varying pressure is given by:

$$p(\bar{r},t) = \mathrm{Re}\left[p(\bar{r})e^{j\omega t}\right] \tag{7}$$

## 5.2 Harmonic Response Analyses

The objective of harmonic analyses is to calculate response of system as a function of frequency based on volumetric flow rate or pressure excitation. Harmonic analysis is a technique used to determine the response of a linear structure to loads that vary sinusoidally (harmonically) with time. The wave equation resolved in acoustic simulation requires mass density $\rho$ and sound velocity $c$ of the fluid media [5].

In harmonic response analyses the following equation is resolved for pure acoustic problems:

$$\left(-\omega^2[M_a] + j\omega[C_a] + [K_a]\right)\{p\} = \{f_F\} \tag{8}$$

$[Ma]$—acoustic fluid mass matrix, $[Ca]$—acoustic fluid damping matrix, $[Ka]$—acoustic fluid stiffness matrix, $\{p\}$—acoustic pressure, $\{ff\}$—acoustic fluid load vector.

## 5.3 Wave Absorption Conditions (Surface of Model)

The exterior structural acoustics problem typically involves a structure submerged in an infinite, homogeneous, inviscid fluid. In FEA we need to truncate the domain. Wave absorption conditions allow us to model a smaller portion of the domain and assume that outgoing waves keep propagation outwards and do not reflect back. For that we use Perfectly Matched Layers (PMLs). PMLs are artificial anisotropic materials that absorb all incoming waves without any reflections, except for the gazing wave that parallels the PML interface in the propagation direction. PMLs are currently constructed by the propagation of acoustic waves in the media for harmonic response analysis [5].

In harmonic analysis the governing lossless and source-free momentum and conservation are given by:

$$\nabla p = -j\omega p_0 \bar{v} \quad \nabla p = -\frac{j\omega p_0}{p_0 c^2} \tag{9}$$

All the above formulas were used for subsequent models as boundary conditions or formulas for conduct acoustic variables in the model.

## 5.4    Models of Acoustic Antiphase

Modeling of acoustic problem was performed by using ANSYS software. On the (Fig. 1) you can see the model of transformer surrounded by air and it generates noise. Acoustic noise is applied as a basic loading of model by pressure condition with value of p = 1 Pa ≈ 90 dB. For basic boundary condition on the outer surface of model we used Perfectly Matched Layers which is not actually shown on this figure. Transformer is located on the ground so there is also applied pressure condition p = 0 Pa to make reflection surface. Harmonic analysis with nominal frequency f = 100 Hz was used for simulation of noise canceling in all models. One part of outputs is presented in the SPL = Sound Pressure Level and unit of contour scales on these models is in dB, second part is presented in Pascals for clear understanding of solution outputs (Fig. 6).

Figure 7 shows the same situation like in previous picture. Scale is set to be in Pa. There is very thin layer with very high level of noise located close to outer surface of transformer. With increasing of distance from transformer the noise level decrease very rapidly close to zero. Due to reflection of the ground the acoustic pressure flows essentially to the top parts of model and also propagates the sides.

On Fig. 8 you can see significant effects of prototype of Antiphase Eliminator which is acting on the transformer. The both elements radiate the pressure wave against each other. The Antiphase Eliminator produces sound wave of the same pressure intensity I = 90 dB but in opposite phase. On this picture you can also see how the pressure wave from the antiphase eliminator deformate the pressure wave from the transformer. The both waves are summed together and the green color which is located close the middle of mutual distance between transformer and eliminator shows the effect of acoustic short circuit.

Figure 9 shows the same situation but scale is in Pascals. There is also zone with insignificant value of pressure which represents place of acoustic short circuit.

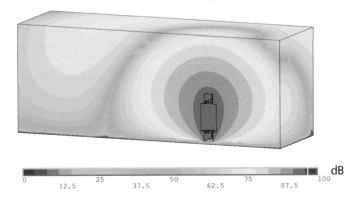

**Fig. 6** Model of the spread of parasitic noise of the transformers

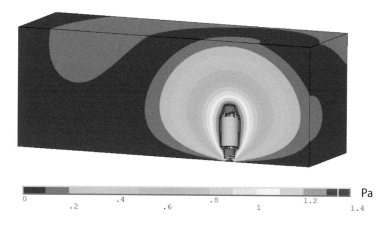

**Fig. 7** Model of the spread of parasitic noise of the transformers in Pa

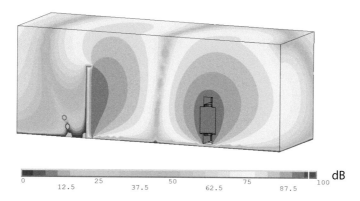

**Fig. 8** Model of antiphase wave action

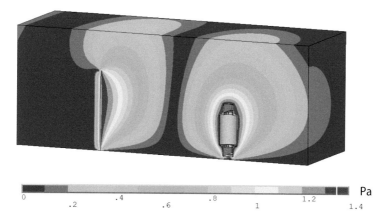

**Fig. 9** Model of antiphase wave action in Pa

Presented models show how can be also noise canceling usefull in case of power electrotechnics. Both models were created with high degree of simplification but as you can see they work very fine. There is also option to change all geometry dimensions and loading values of model to simulate our problem more close to reality. Also they are used for researching of the new generation of prototype of antiphase eliminator. Using of acoustic numerical models and their comparison with real acoustic measurement is also useful due to immediate evaluation of changes in the real time. From this models are clearly seen that the our Antiphase Eliminator is functional device.

# 6 Conclusion

The aim of this article was to help other researchers of this problem to define the conditions for modeling antiphase acoustic short. A decision was made to continue with the research on the basis of the extremely successful completion of the first phase of the research, wherein the functional prototype of the Antiphase Eliminator was created. The next step will involve the actual digitalization of the instrument and the overall optimization of the method. The optimization will consist of a simplified, for the current time, analogical chain providing the process of treating the actual parasitic sound signal. An algorithm has been created from the acoustic equations which will make up the mathematical basis for adding up the two acoustic signals. A completely new prototype created on the basis of this process will be more economical, more effective and which will have a wider field of application.

**Acknowledgments** This paper has been elaborated in the framework of the project New creative teams in priorities of scientific research, reg. no. CZ.1.07/2.3.00/30.0055, supported by Operational Programme Education for Competitiveness and co-financed by the European Social Fund and the state budget of the Czech Republic and supported by the IT4 Innovations Centre of Excellence project CZ.1.05/1.1.00/02.0070, funded by the European Regional Development Fund and the national budget of the Czech Republic via the Research and Development for Innovations Operational Programme. This paper was conducted within the project ENET— CZ.1.05/2.1.00/03.0069, LO1404: Sustainable development of ENET Centre. Students Grant Competition project reg. no. SP2015/170, SP2015/178, project LE13011 Creation of a PROGRES 3 Consortium Office to Support Cross-Border Cooperation (CZ.1.07/2.3.00/20.0075) and project TACR: TH01020426.

# References

1. Pokorny, V., Misak, S.: Using the antiphase for elimination the noise of electrical power equipment. AEEE Adv. Electr. Electron. Eng. **11**(3) (2013). doi:10.15598/aeee.v11i3.794
2. Chu, M.: http://www.gilmore2.chem.northwestern.edu/tech/anr_tech.htm (2001). Accessed 19 Apr 2010
3. Tomlinson, H.: Anti-noise by Tomlinson Holman. http://www.mefeedia.com/ (2009). Accessed 15 May 2010

4. Hinamoto, Y., Sakai, H.: A filtered-xLMS algorithm for sinusoidal reference signals—effects of frequency mismatch. IEEE Sign. Process. Lett. **14**, 259–262 (2007). doi:10.1109/LSP.2006.884901
5. ANSYS Theory Manual (2015)

# Index

Printed in the United States
By Bookmasters